USS *CONSTITUTION*
A MIDSHIPMAN'S POCKET MANUAL, 1814

Compiled and introduced by Eric L. Clements

Osprey Publishing
c/o Bloomsbury Publishing Plc
PO Box 883, Oxford, OX1 9PL, UK

c/o Bloomsbury Publishing Inc.
1385 Broadway, 5th Floor, New York, NY 10018, USA
E-mail: info@ospreypublishing.com

www.ospreypublishing.com

OSPREY is a trademark of Osprey Publishing Ltd, a division of Bloomsbury Publishing Plc.

First published in Great Britain in 2017

ISBN: HB: 978-1-4728-2793-7
ePub: 978-1-4728-2792-0
ePDF: 978-1-4728-2791-3
XML: 978-1-4728-2794-4

17 18 19 20 21 10 9 8 7 6 5 4 3 2 1

Typeset in Adobe Caslon Pro by Deanta Global Publishing Services, Chennai, India
Printed in China through World Print Ltd.

Front Cover: USS *Constitution* © H. Armstrong Roberts / Classic Stock / Getty Images

Osprey Publishing supports the Woodland Trust, the UK's leading woodland conservation charity.
Between 2014 and 2018 our donations are being spent on their Centenary Woods project in the UK.

To find out more about our authors and books visit www.ospreypublishing.com.
Here you will find extracts, author interviews, details of forthcoming events and
the option to sign up for our newsletter.

CONTENTS

INTRODUCTION

"The tenure of a sailor's existence is certainly more precarious than any other man's, a soldier's not excepted. Who would not be a sailor? I, for one."

—*Surgeon Amos A. Evans*, USS *Constitution*, 1812

What follows is a book that was never written. In 1814, two years into the War of 1812, the US Navy Department published its revised naval regulations. What if the department, concerned about officer training in the middle of a war, had published a guide for midshipmen at the same time? To extend this fantasy further, what if the department had issued these guides tailored to the classes of vessels or even to the individual ships then in its service? That is the premise of this book.

THE U.S. NAVY IN THE FEDERAL ERA

After ratifying the Treaty of Paris of 1783, ending its Revolutionary War, the United States disbanded its navy, selling off the last of those ships in 1785. The U.S. Navy would be resurrected due to the activities of the Barbary pirates in the Mediterranean. Deprived the aegis of the Royal Navy upon independence, the new country's merchant vessels became vulnerable to the attacks of the Barbary corsairs operating out of Tripoli, Tunis, and Algiers. The corsairs captured American ships and cargoes and held their crews for ransom. In 1794 Congress authorized the creation of a navy and the construction of six ships, among them USS *Constitution*, to address the Mediterranean situation.

Beginning in early 1793 the United States faced another threat to its seaborne trade. The wars of the French Revolution and Napoleonic era would last, with only brief intermissions, from 1792–1815. The United States would attempt to remain neutral during this long conflict and to continue to trade with both Great Britain and France, a posture satisfactory to neither European power. The conflict produced by these circumstances would ultimately lead the United States to declare war on Great Britain in 1812, the war in which USS *Constitution* would win her renown.

THE BARBARY CORSAIRS

At the end of March 1794, having decided that "the depredations committed by the Algerine corsairs on the commerce of the United States render it necessary that a naval force should be provided for its protection," the U.S. Congress passed "An Act to Provide a Naval Armament." This authorized the construction and manning of four 44-gun and two of 36-gun frigates, *Constitution* being one of the former vessels.

With its fleet still under construction, however, the United States at first had to negotiate rather than fight. In September 1795 the U.S. signed a treaty with the Dey of Algiers in which it agreed to pay $1 million to ransom 155 American sailors, and to make annual tribute payments in exchange for an end to corsairs' attacks on American shipping in the Mediterranean and off the coasts of Spain and Portugal in the Atlantic. Treaties extending similar tributes followed in November 1796 with Tripoli and in August 1797 with Tunis.

THE UNITED STATES AND THE WARS OF THE FRENCH REVOLUTION

In the meantime, the United States confronted another, more powerful foe much closer to home. In February 1793, after the execution of Louis XVI, France went to war against Great Britain, Spain, and the Netherlands. In response, on April 22 President George Washington issued a proclamation of neutrality, declaring that "the duty and interest of the United States require, that they should with sincerity and good faith adopt and pursue a conduct friendly and impartial toward the belligerent Powers."

The United States would attempt to maintain that policy throughout the wars of the French Revolution and Napoleonic era. The belligerents, however, had little interest in upholding American neutrality and from the start both sides interfered with American trade intended for their adversary. In May 1793 the French government ordered its navy to seize neutral ships carrying supplies to British, Spanish, or Dutch ports. The following month Great Britain countered with orders to seize neutral ships, including American, headed to French ports. This war spread to the Western Hemisphere at the end of the year thanks to a British Order in Council, issued that November, to seize any neutral vessel carrying exports from French colonies in the Caribbean, which led to the capture of American ships and the impressment of their crews.

THE QUASI-WAR, 1796–1800

The Washington administration and later that of John Adams attempted to negotiate U.S. differences with the European powers, but soon discovered that any settlement with one power only provoked retaliation from its opponent. In 1795 the U.S. reached a modest accommodation with Great Britain through Jay's Treaty, only to draw the ire of France, which suspended diplomatic relations with the U.S. at the end of 1796.

In May 1797 President Adams dispatched a three-member commission to Paris to negotiate settlement of the differences with France, but his action only exacerbated the crisis. In October representatives of the French minister of foreign affairs attempted to bribe the U.S. commissioners, seeking payments as a precondition to negotiations. When Adams released details of what came to be called the XYZ Affair to Congress the following April, public outrage led Congress to suspend commerce with France and its dependencies in June 1798 and the following month to renounce the 1778 treaties of alliance and commerce with France. What followed was a two-year-long, undeclared naval war in the Caribbean between French privateers and the new ships of the U.S. Navy. This Quasi-War was settled by the Treaty of Mortefontaine, in which France and the U.S. agreed to resume normal diplomatic relations, end the Quasi-War, and annul the treaties of 1778.

THE FIRST BARBARY WAR, 1801–05

Scarcely had the conflict with France in the Caribbean been resolved before trouble developed again in the Mediterranean. In May 1801 the Pasha of Tripoli, whose demands for greater tribute remained unmet, declared war on the United States. Rather than continue to pay tribute, the new U.S. president Thomas Jefferson, with the naval means at his disposal, decided to challenge the Barbary corsairs in what became the First Barbary War. Five years later, in June 1805, the American campaign culminated in a treaty between the United States and Tripoli, in which the United States agreed to pay a $60,000 ransom to recover the crew of the captured U.S. frigate *Philadelphia*, but in return Tripoli renounced further privateering or tribute.

EARLY SERVICE OF USS *CONSTITUTION*

USS *Constitution* entered service in July 1798 at the beginning of the Quasi-War. She made two undistinguished patrols during that conflict, then was retained in inactive reserve, "in ordinary," at Boston from 1801 to 1803. In August 1803 she sailed for the Mediterranean and service in the First Barbary War. She remained in that sea for four years, participated in five bombardments of Tripoli in August and September 1804 as Commodore Edward Preble's flagship, and the draft of the treaty with Tripoli ending the war was completed aboard *Constitution* on 3 June 1805 under Commodore John Rodgers. After her return to the United States in October 1807 *Constitution* was inactive until being assigned to coastal patrols in August 1809. Her last peacetime cruise before the War of 1812 was a diplomatic mission to France, Holland, and England from August 1811 to February 1812. Her call at Portsmouth, England, came in November 1811, amidst the Anglo-American tensions that would lead to war only seven months later.

THE NAPOLEONIC WARS, 1803–15

By the time the issues had been settled in the Mediterranean with the Barbary States, European hostilities had resumed after a short respite brought about by the Treaty of Aliens of March 1802. This second phase of the French and Allied conflict lasted from the resumption of hostilities between Britain and France in May 1803 until Napoleon's final defeat at Waterloo in June 1815.

Once again, the U.S. and its trade were caught between the combatants in a crossfire of decrees. In May 1806 Great Britain imposed a blockade on the northwestern coast of Europe. In response, Napoleon issued the Berlin Decree that November. This established the Continental System, which forbid any country allied to or dependent upon France to trade with Britain. The British government responded with the Orders in Council of November 1807. These forbid French trade with the United Kingdom, its allies, or neutrals, and required neutral ships to stop at an English port for inspection prior to sailing for the Continent. Napoleon countered with the Milan Decree of December 1807, declaring that any ship, including neutral vessels, that called at a British port or permitted inspection by the Royal Navy would be considered British and therefore subject to seizure and confiscation. Thus, if U.S. ships followed the Orders in Council they were liable to French seizure under the Milan Decree; if they adhered to the terms of the Milan Decree, they were liable to British seizure under the Orders in Council.

"FREE TRADE & SAILORS' RIGHTS"

That same year, this war of decrees concerning neutral rights became a shooting war over impressment on the opposite side of the Atlantic. On 22 June 1807 the 50-gun frigate HMS *Leopard* accosted the 36-gun U.S. frigate *Chesapeake* off Norfolk, Virginia, demanding to board the ship to recover Royal Navy deserters. When *Chesapeake*'s commander refused Leopard opened fire, killing three Americans and wounding eighteen. After *Chesapeake* struck her colors *Leopard* removed four of her crewmen, three of them American born.

Intense outrage in the U.S. over the Chesapeake-Leopard incident led Congress to pass the Embargo Act in December 1807. This attempted to force both Britain and France to recognize U.S. neutral rights by forbidding American ships to sail for foreign ports and foreign ships from loading cargoes in American ports. Although in the long term American trade embargoes would damage the British economy, in the short term the Embargo Act, passed by a Republican Congress, most injured the maritime trade and economy of Federalist New England. The region's opposition to trade embargoes and to anti-British policies in general would lead New Englanders to smuggling with British Canada and later to opposing the War of 1812.

On 1 March 1809 Congress replaced the ineffective and unpopular Embargo Act with the Non-Intercourse Act, which reopened overseas commerce with all nations except Britain and France, but empowered the president to reinstate trade with either country if it renounced its violations of U.S. neutral rights. Fourteen months later, on 1 May 1810 Congress replaced the Non-Intercourse Act with the even weaker and more convoluted Macon's Bill No. 2. This lifted the embargoes yet authorized the president to re-impose the embargo on either Britain or France if the other power agreed to respect U.S. neutral rights. With that, the French government pulled a ruse, implying that the Berlin and Milan decrees had been revoked. Believing this to be so, Madison re-imposed the embargo on Great Britain in November 1810 further straining

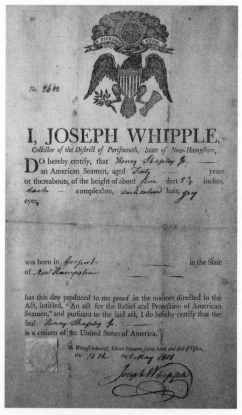

© USS Constitution Museum Collection.

Anglo-American relations. Britain reacted by stationing warships off New York and by continuing to impress American sailors.

Even after France's trick became known to Madison in September 1811, he retained the embargo on Great Britain. Great Britain was seizing more American ships than France, impressing American sailors, and, allegedly, continuing to incite Native American attacks upon American settlers in the northwest. When Madison sent his war message to the U.S. Congress in June 1812 western issues received only brief mention in his catalog of Great Britain's "series of acts hostile to the United States as an independent and neutral nation." Most of his war message was devoted to denouncing Britain's attacks on American neutral trade and its impressment of U.S. citizens. Congress responded on 18 June with a declaration of war, voting a party-line approval, 19 to 13 in the Senate and 79 to 49 in the House. The declaration was supported heartily in the agrarian South and West, and hardly supported at all in the commercial and maritime Northeast.

MR. MADISON'S WAR

With the exception of landlocked Vermont, every New England state opposed the war, as did the maritime states of New York, New Jersey, and Delaware. In Boston "A New England Farmer" published "Mr. Madison's War," protesting "an offensive and ruinous war against Great Britain." New England objected officially as well. A week after the declaration of war the governor of Massachusetts declared a statewide fast in protest and the state's legislature issued a proclamation declaring the war to be against the public interest and promising to provide military forces only for defense. Both Connecticut and Massachusetts would refuse to provide militia for federal service later that summer. While President Madison won reelection comfortably that November, Federalists doubled their strength in Congress. In December 1813 Madison would ask Congress for a new embargo act forbidding any trade with the British to stop New England's trading with British Canada.

PEACE FEELERS

On 16 June 1812 two days before the U.S. declared war, Lord Castlereagh, Britain's Secretary of State for Foreign Affairs, announced a suspension of Britain's Orders in Council concerning neutral shipping. The British economy was suffering the economic effects of both the American and Continental embargoes. Both Britain

Mr. Madison's War.

A

DISPASSIONATE INQUIRY

INTO THE

REASONS ALLEGED BY MR. MADISON

FOR DECLARING AN

OFFENSIVE AND RUINOUS WAR

AGAINST GREAT-BRITAIN.

TOGETHER WITH

SOME SUGGESTIONS

AS TO A

PEACEABLE AND CONSTITUTIONAL MODE

OF AVERTING THAT DREADFUL CALAMITY.

BY A NEW-ENGLAND FARMER.

"Poor is his triumph, and disgrac'd *his* name,
Who draws the sword for empire, wealth, or fame:
And poorer still their statesmen's share of praise,
Who at a tyrant's nod their country's standard raise:
For them though wealth be blown on every wind,
Though *twice ten* nations crouch beneath their blade,
Virtue disowns them, and their glories fade.
For them no prayers are pour'd, no pæans sung,
No blessings chaunted from a nation's tongue;
Blood marks the path to their untimely bier:
The curse of orphans and the widow's tear,
Cry to high Heaven for vengeance on their head,
Alive deserted, and accurs'd when dead."

BOSTON:
PRINTED BY RUSSELL & CUTLER.
1812.

and the United States began to seek a compromise almost as soon as the war began. Proposals went back and forth in the summer and fall of 1812 but nothing came of these efforts until November 1813, when Castlereagh sent Madison a letter offering direct negotiations. Madison accepted immediately and named a five-member peace commission to represent the United States. Peace negotiations opened at Ghent, in northwest Belgium, on 8 August 1814.

THE U.S. NAVY IN THE WAR OF 1812

In 1812 Great Britain possessed a 600-ship navy, manned by 130,000 sailors, with about 80 of those ships active in the Western Hemisphere, deployed from Newfoundland to Jamaica. At the beginning of 1812 the U.S. Navy mustered an active force of 14 seagoing ships, its three 44-gun frigates being the heaviest, and about 1,000 sailors. Any American successes would thus be limited to commerce raiding or winning single-ship encounters. Nothing the U.S. would or could do could affect Great Britain's strategic supremacy at sea. Nevertheless, tactical opportunities occasionally presented themselves, and the U.S. Navy's victories were important both to the service itself and to the country at large.

The first American success came eight weeks into the war, on 13 August, when the 32-gun frigate USS *Essex* captured the 20-gun sloop HMS *Alert*. Six days later *Constitution* scored the first of her three triumphs, defeating the frigate HMS *Guerriere*. The U.S. earned two more victories in October, when the 18-gun sloop USS *Wasp* defeated the 18-gun brig HMS *Frolic* on October 17 and *Constitution*'s sister ship, the 44-gun *United States*, bested the 38-gun HMS *Macedonian* on October 25. The year ended with *Constitution*'s second victory, the defeat of HMS *Java* on 29 December.

The battle with *Guerriere* seems to have been the origin of *Constitution*'s nickname, "Old Ironsides". The story goes that a sailor on the *Guerriere* saw 18-pound British cannonballs bouncing off the hull of *Constitution* and exclaimed, "Huzzah, her sides are made of iron!"

By 1813 the novelty of American victories began to fade and the weight of the Royal Navy began to tell. While the U.S. scored a few triumphs that year, notably the 18-gun sloop USS *Hornet*'s defeat of the 20-gun sloop HMS *Peacock* in February and Captain Oliver Hazard Perry's defeat of the British fleet on Lake Erie, the navy also suffered its first significant defeat that June when the 50-gun frigate HMS *Shannon* captured the 36-gun frigate USS *Chesapeake*. Worse, from the American point of view, the Royal Navy was tightening its blockade of the U.S. coast and using its supremacy to strike at will along that coast, raiding points in Maine, Massachusetts, Connecticut, and the Chesapeake Bay.

THE YEAR 1814

In 1814 (the year of the pretended first edition of this pocket manual), the war situation became worrisome for the United States. The Royal Navy extended its blockade along the entire U.S. Atlantic Coast, which the comparatively miniscule U.S. Navy could do nothing to prevent. With American overseas trade and thus tariff revenues from imports substantially reduced, the United States government defaulted on the national debt in November. Meanwhile, dissatisfaction with the war increased in New England.

While American privateers were enjoying considerable success at the beginning of March 1814 the U.S. Navy had a total of nine ships at sea, only three of them mounting more than twenty guns. Within a month, the largest, the 44-gun *Constitution*, would return to port after an abbreviated patrol, while the next in size, the 32-gun *Essex*, would be captured by two British warships off Valparaiso, Chile. At the time, the U.S. had three 44-gun and three 36-gun frigates in operation, as well as an additional three 44s and three 74-gun ships of the line under construction; however Napoleon's defeat, abdication, and exile to Elba in April permitted Britain to shift its attention and significant naval and military resources to its war against the United States.

The American nadir came after the Royal Navy disembarked an army of 4,000 veteran soldiers under General Robert Ross at Benedict, Maryland, on 19 August. Ross' army routed a patchwork American force half-again its size at Bladensburg, Maryland, on 24 August then captured and burned Washington, D.C., unopposed, the same evening. Although the Americans successfully defended Baltimore three weeks later, an end to the war seemed as remote as ever.

USS CONSTITUTION *IN 1814*

Constitution departed Boston for a six-month war patrol on 30 December 1813. Her cruise in the Caribbean yielded four prizes in a week, but also ended ingloriously on 3 April 1814 due to an outbreak of scurvy and a cracked mainmast, with *Constitution* being chased into harbor at Marblehead, Massachusetts, by the frigates HMS *Junon* and HMS *Tenedos*. Shifted back to Boston two weeks later, *Constitution* would remain there, blockaded, for the following eight months. A board of inquiry investigated *Constitution*'s captain, Charles Stewart, for curtailing his patrol, but he was retained in command, which proved fortunate. On 17 December after bad weather drove away the British blockading ships *Constitution* put to sea for her last war patrol and last hostile encounters with the Royal Navy.

MIDSHIPMEN IN THE FEDERAL-ERA NAVY

Midshipmen, the lowest ranking apprentice navel officers learning their trade, would have had some formal education before receiving their appointments. A few had even attended college. Their academic education would generally continue aboard navy ships under the direction of the sailing master professionally, and the chaplain or school teacher scholastically. Midshipmen also kept watch, manned stations at quarters, usually in charge of one of the ship's gun pairs, commanded boats, shore parties, or other details, and served as masters of the prizes their ship captured.

Although the midshipmen of the early years of the Republic ranged from pre-teens to men in their mid-twenties, the average midshipman at appointment was in his late teens (the median age being seventeen and a half), and considered an adult by the standards of his era. Seven in ten new appointees were fifteen to twenty years old. In 1814 they came mostly from the Mid-Atlantic states, followed by the New England states, and the Virginia-Maryland-Washington, D.C., area, perhaps three-quarters of them from the seaboard areas of those places. Midshipmen were supposed to be from "respectable" backgrounds, with a modest middle-class status generally sufficient.

Before the advent of the U.S. Naval Academy in 1845 American midshipmen trained entirely aboard ships at sea. Then, as now, appointment as a midshipman was highly competitive. In February 1814, during a war, Secretary of the Navy William Jones wrote of there being ten applicants for every vacancy in the navy's officer corps. A few men became naval officers through the position of sailing master, a warship's navigator, having come to that situation from previous experience as captains or mates in merchant ships; but most of those who eventually gained a lieutenancy started as midshipmen.

SOME CAVEATS

The idea of this pocket manual is to tell something of the design and history of, and life aboard USS *Constitution* during the War of 1812 by casting the reader in the role of a midshipman reporting aboard *Constitution* in the fall of 1814, just before she departed on her last war patrol under Captain Charles Stewart. Neophytes also enlisted aboard *Constitution* as "landsmen" and ship's boys. As a former enlisted man myself, I might have written a guide for ordinary sailors, a War of 1812 version of the navy's *Bluejacket's Manual* first published in 1902. The difficulty with that approach is that most enlisted sailors in 1812 were illiterate. As an officer apprentice and at least nominally a gentleman, almost all midshipmen had at least some ability to read.

On the subject of literacy, however, it should be understood that those writing the original documents that follow were mostly men of the sea rather than of letters. Given that, the unevenness of available education and the inconsistency of literary standards of grammar and punctuation two centuries ago, the quality of the writing they produced varied considerably. In an effort to provide more clarity, I have modernized capitalization, punctuation, and some spellings in cases where doing so did not alter the original meaning.

In the spirit of the project, and to create transitions between the original sources, I have attempted, however successfully, to mimic the style of writing of 1812. My intention generally is to introduce and blend the sources that form the bulk of each section while providing them with context and background. To separate factual sources and fictional narration, my narration appears in italics. To avoid the encumbrance of numbered notation, which would also betray the pretense, the sources used in each section are credited generally at the end of the book.

As to the sources themselves, information about specific facets of daily life two centuries ago is sometimes difficult to obtain. Therefore we must sometimes deal in generalities and use sources that do not directly apply to USS *Constitution* in 1814. Thus the captain's orders cited are from eleven years earlier and the medical text from thirty-two years later. While it would be preferable if more exact sources existed, a useful sense of the era may nevertheless be derived from the materials that have survived.

Finally, two limitations are inherent in writing a wartime pocket manual supposedly intended to train and exhort American midshipmen: It is necessarily partisan and celebratory in tone, omitting any balancing British perspective.

It also obscures the costs of battle, men horribly killed and maimed. Some sense of that price paid appears in the "addenda" included in this fiction in a second printing after the end of the war. That section also details some of the adventures the reader-midshipman would have experienced on his first war patrol and *Constitution*'s last.

CONSTITUTION'S SUBSEQUENT SERVICE

By the time *Constitution* returned to the U.S. from her last war patrol, the Treaty of Ghent ending the War of 1812 had been ratified by the United States Senate in February 1815. After *Constitution*'s return to Boston at the end of May she was placed in ordinary, from January 1816 until February 1821. Reactivated, she sailed for the Mediterranean in May 1821 and, with the exception of the summer of 1824, would spend the next seven years in the U.S. Navy's Mediterranean squadron.

Six more years in ordinary followed, starting in mid-1828, before *Constitution* was reactivated for more Mediterranean service. That period in ordinary ended with another first for *Constitution*. On 24 June 1833 she became the first ship to enter and be served by the Charleston (Massachusetts) Navy Yard's new Dry Dock No. 1, where she is maintained to the present.

Constitution spent the years from August 1835 to August 1838 as flagship of the Mediterranean squadron, followed by a turn as flagship of the U.S. Pacific squadron cruising off the west coast of South America, a duty that required her to sail around Cape Horn in September 1839 and July 1841. She returned from the Pacific station in the fall of 1841 to Norfolk, Virginia, where *Constitution* was inactive for the following two and a half years.

Constitution's next voyage, which departed New York on 29 May 1844, circled the globe. She called at the Azores and Rio de Janeiro before passing the Cape of Good Hope and visiting Madagascar. From there she sailed north, stopping at ports on the east coast of Africa, before visiting Sumatra, Singapore, Borneo, Brunei, Indochina, China, and the Philippines. A fifty-day passage from Manila brought *Constitution* to Honolulu, where she was visited by King Kamehameha III before sailing for Mexico. *Constitution* left Mazatlán three days before the beginning of the Mexican-American war. The ship returned to Boston on 27 September 1846 via Valparaiso, Cape Horn, and Rio de Janeiro after a voyage of 28 months and 52,370 miles.

The old frigate, over fifty and soon to be outmoded by iron-hulled steamships, had two voyages remaining in her active career. The first was another Mediterranean deployment from December 1848 to January 1851, which included the following unusual log entry:

2ⁿᵈ Day of August 1849. . . . From 8 to meridian light breezes from the N & E & pleasant[;] at 11.15 his holiness Pope Pius IX and his majesty Ferdinand II king of the two Sicilies visited the ship[;] manned the yards and saluted [them] with 21 guns on their arrival on board and also on their departure from the ship[,] the vessels of war in the harbour saluting on both occasions.

These remarks recorded the first visit by a pontiff to U.S. territory.

Constitution's final active deployment, of twenty-seven months and 42,166 miles, began in March 1853. It took her first on a final circuit of the Mediterranean before she sailed for the west coast of Africa to join the U.S. Navy's Africa squadron organized to intercept ships engaged in the illegal international slave trade. Here *Constitution* scored her last victory, capturing the American slaver *H. N. Gambrill* out of New York. *Constitution* reached Portsmouth, New Hampshire, on 2 June 1855 ending her operational career of 57 years.

Constitution spent the following five decades in a variety of roles, first as a school ship in various capacities, starting with the Naval Academy. After being declared unfit for further sea service in 1881, she became a receiving ship at Portsmouth, New Hampshire, with her masts cut down and a barracks built over her spar deck. *Constitution* was rescued from that undignified role by her own centennial, being towed to Boston in the summer of 1897.

Following her centennial celebration *Constitution* was moored in a backwater at the Charleston Navy Yard and left to decay. That decay was disguised by a renovation in 1907 in which "she was repaired above the waterline to a certain extent by listing, but little work was accomplished at the two ends of the ship. The old roof structure was removed; new masts, spars, tops, etc., were made at the Portsmouth Navy Yard and the ship masted and sparred at Boston. A portion of the standing and running rigging was installed."

A survey conducted in 1924, however, found Old Ironsides almost ruined with significant rot and distortion in her hull. The survey report "concluded that if the CONSTITUTION was not rebuilt in the immediate future the vessel would go to ruin and it would be lost to the nation." The first problem was to raise the money necessary to rebuild her, estimated originally at $400,000. The next problem was even getting *Constitution* into Dry Dock No. 1, "the hull structure [being] so weak, distorted and hogged that there was a possibility of the two ends falling off."

After sufficient funds were secured, in large part through a nationwide subscription campaign, Old Ironsides entered the dock for the first time in thirty years on 16 June 1927 emerging rebuilt on 15 March 1930. Repairs to rigging and interiors took another year, with the entire restoration eventually costing $921,108. In summarizing the project, a navy report recorded that "the remark has been made that a new ship of the type could have been built as cheaply, which might be true."

Recommissioned on 1 July 1931 *Constitution* departed Boston the following day, towed by the minesweeper USS *Grebe* to begin her "National Cruise" of East, Gulf, and West Coast ports from Maine to Puget Sound. *Constitution*'s last epic voyage ended back in Boston on 7 May 1934 after the ship had traveled over 22,000 miles, visited 90 ports, and hosted more than 4.6 million visitors.

The following fifty years passed with three significant milestones. The first was an Act of Congress in 1954 authorizing the secretary of the navy "to repair, equip and restore the United States ship *Constitution*, as far as may be practicable, to her original condition, but not for active service, and thereafter to maintain [the ship] at Boston, Massachusetts." The second was the closing of the Charlestown Navy Yard in 1974 and the creation that same year of the Boston National Historical Park, which includes a significant portion of the old yard and a number of its buildings, Dry Dock No. 1, and *Constitution*'s regular berth. The third, concurrent with the establishment of the park, was a restoration of the ship to her 1812 appearance in anticipation of the U.S. Bicentennial, and the simultaneous creation of the USS *Constitution* Museum. The museum opened in the former dry dock machinery building in April 1976 and the ship's restoration was completed on 1 July. The Bicentennial celebrations culminated with a visit to Boston harbor by HMY *Britannia* and a visit to USS *Constitution* by Queen Elizabeth II and Prince Philip on 11 July 1976.

A number of the ship's traditions started that year, including the retention of a permanent force of artisans to see to her maintenance and the firing of morning and evening guns. The annual Independence Day turnaround cruise began the following year, done to permit the ship to weather more evenly. The ship was dry docked again for significant overhauls from September 1992 to September 1995 when replicas of her original diagonal riders were installed, and again from 2015 to 2017 for repairs, restorations, and re-coppering.

AND A VERY HAPPY ENDING

And what became of those warring nations of two centuries before? As the Queen's visit attests, relations between Great Britain, Canada, and the United States have improved significantly, and they did so very quickly. While the Treaty of Ghent did little but stop the war, with Napoleon defeated and the Pax Britannica ascendant, the most significant issues that drove the two nations to war, neutral rights and impressment, ceased to matter. The two countries also soon settled their territorial questions. Representatives of Great Britain and the United States signed the Convention of 1818, which reaffirmed the Treaty of Ghent and fixed the boundary between British Canada and the United States at the forty-ninth parallel between the Lake of the Woods and the Continental Divide, with the territory to the west open to joint settlement.

The previous year, in the Rush-Bagot Treaty of April 1817, U.S. Acting Secretary of State Richard Rush and the British Ambassador to Washington Sir Charles Bagot resolved

that the naval force to be maintained upon the [Great] Lakes by the United States and Great Britain shall, henceforth, be confined to the following vessels on each side, that is: On Lake Ontario to one vessel not exceeding One Hundred Tons burden, and armed with one eighteen-pound cannon. On the Upper Lakes to two vessels not exceeding the like burden each, and armed with like force, and on the waters of Lake Champlain to one vessel not exceeding like burden and armed with like force.

This understanding remains in effect today: the oldest arms-control agreement extant, on the longest undefended border on the planet.

MISTER MIDSHIPMAN

Navy Dept.
19 June 1812

By the enclosed papers you will perceive that War has been declared [against] the United Kingdom of Great Britain & Ireland and her dependencies—and that the President has been authorized to use the whole land and naval force of the United States to carry the same into effect.

In virtue of this authority, you will consider yourself, and all the officers and vessels under your command as having every belligerent right of attack, capture & defense. Be upon your guard. Instructions more in detail will be shortly transmitted to you.

P. Hamilton

So, too, you, Mister Midshipman. Instructions more in detail will be transmitted to you in the following chapters, but suffice for the moment to say that Navy Secretary Hamilton's orders apply to you as surely as they do to our navy's captains.

Be upon your guard.
Be obedient to your superiors.
Be devoted to your duties.
Be faithful to your Oath.

A MIDSHIPMAN'S OATH OF ALLEGIANCE

I [name and home town] appointed [a midshipman in the navy of the United States] do solemnly swear to bear true allegiance to the United States of America, and to serve them honestly and faithfully against all their enemies or oppressors whomsoever; and to observe and obey the orders of the President of the United States of America, and the orders of the officers appointed over me, and in all things to conform myself to the rules and regulations which now are or hereafter may be directed, and to the articles of war which may be enacted by Congress, for the better government of the navy of the United States, and that I will support the constitution of the United States.

SWORN BEFORE ME   [signed by oath taker]
[date]
[signed by witness]

A MIDSHIPMAN'S DUTIES

Of the duties of Midshipmen.

No particular duties can be assigned to this class of officers.

They are promptly and faithfully to execute all the orders for the public service, of their commanding officers

The commanding officers will consider the midshipmen as a class of officers meriting, in an especial degree, their fostering care. They will see, therefore, that the schoolmasters perform their duty towards them, by diligently and faithfully instructing them in those sciences appertaining to their departments; that they use their utmost care to render them proficient therein.

Midshipmen are to keep regular journals, and deliver them to the commanding officer at the stated periods, in due form.

They are to consider it as the duty they owe to their country, to employ a due portion of their time in the study of naval tactics, and in acquiring a thorough and extensive knowledge of all the various duties to be performed on board of a ship of war.

And:

. . . Of the duties of a Chaplain: . . . He shall perform the duty of a schoolmaster; and to that end, he shall instruct the midshipmen and volunteers in writing, arithmetic and navigation, and in whatsoever may contribute to render them proficient. . . .

—From "Naval Regulations Issued by Command of the President of the United States of America," 1814.

THE KEEPING OF JOURNALS

Regulations respecting the form and mode of keeping the Log-book and Journals on board of ships, or other vessels, of the United States.

For the purpose of establishing uniformity, the President orders as follows, viz.

1. The quarter-bill, log-tables or book, and journals of the officers, must be kept conformably to the [appropriate] models.

2. The captains or commanders will cause to be laid before them, the first and fifteenth of every month, the journals of the sea lieutenants, masters, midshipmen, and volunteers under their orders, and will examine and compare them with their own.

3. If any of the said journals contain observations or remarks which may contribute to the improvement of geography, . . . [or] remarks relative to the directions and effects of currents, tides or winds: the officers or persons appointed to examine them, will make extracts of whatever appears to merit to be preserved; and . . . the same shall be . . . transmitted with their opinion thereon to the secretary of the navy, to be preserved in the depot of charts, plans, and journals.

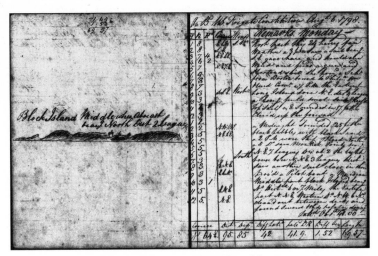

© National Archives of the United States.

A MIDSHIPMAN'S DRESS

The following description of the "Uniform Dress," for the officers of the Navy of the United States, is substituted for that hitherto established; and is to take effect on the first day of January, 1814, to which all officers therein designated, are ordered to conform. . . .

Midshipmen's Full Dress

The Coat of blue cloth, with lining and lapels of the same; the lapels to be short, with six buttons; standing collar, with a diamond formed or gold lace on each side, not exceeding two inches square; with no buttons on the cuffs or pockets.

Pantaloons and Vest white the same as the lieutenant's, except the buttons on the pocket of the vest.

Undress

A short coat, rolled cape, with a button on each side.

A Midshipman, when he acts as lieutenant, by order of the Secretary of the Navy, will assume the uniform of a lieutenant. . . .

Midshipmen, when in full dress, to wear plain cocked hats, half boots, and swords. . . .

A MIDSHIPMAN'S QUARTERS AND MESS

Midshipmen are quartered on the berth deck directly forward of the wardroom, midway between the main- and mizzen masts, in the area known as steerage. You will be assigned to one of the approximately ten-by-ten-foot staterooms in steerage, starboard or larboard, each of which accommodates eight of the ship's sixteen midshipmen. Here, you will sling your hammock and stow your gear in one of the lockers provided. Here also you will eat your meals, served atop the mess chest used to store your eating utensils. More information about the ship's design and arrangement will be found in the following chapter.

A MIDSHIPMAN'S DEPORTMENT

Below are regulations from the Articles of War (An Act for the Better Government of the Navy of the United States, April 23rd, 1800) that pertain to the character and conduct of officers, and thus midshipmen. These you must obey and uphold:

Section 1... Art. 1. The commanders of all ships and vessels of war belonging to the navy, are strictly enjoined and required to show in themselves a good example of virtue, honour, patriotism and subordination; and be vigilant in inspecting the conduct of all such as are placed under their command; and to guard against, and suppress, all dissolute and immoral practices, and to correct all such as are guilty of them, according to the usage of the sea service....

Art. III. Any officer . . . who shall be guilty of oppression, cruelty, fraud, profane swearing, drunkenness, or any other scandalous conduct, tending to the destruction of good morale, shall . . . be cashiered, or suffer such other punishment as a court martial shall adjudge....

Art. IV. Every commander or other officer who shall, upon signal for battle, or on the probability of an engagement, neglect to clear his ship for action, or shall not use his utmost exertions to bring his ship to battle, or shall fail to encourage, in his own person, his inferior officers and men to fight courageously, such offender shall suffer death, or such other punishment as a court martial shall adjudge; or any officer neglecting, on sight of any vessel or vessels of an enemy, to clear his ship for action, shall suffer such punishment as a court martial shall adjudge; and if any person in the navy shall treacherously yield, or pusillanimously cry for quarters, he shall suffer death, on conviction thereof, by a general court marital.

Art. V. Every officer . . . who shall not properly observe the orders of his commanding officer, or shall not use his utmost exertions to carry them into execution, when ordered to prepare for, join in, or when actually engaged in battle; or shall at such time, basely desert his duty or station, either then, or while in sight of an enemy, or shall induced others to do so, .. shall, on conviction thereof by a general court martial, suffer death or such other punishment as the said court shall adjudge....

Art. VI. Every officer . . . who shall through cowardice, negligence, or disaffection in time of action, withdraw from, or keep out of battle, or shall not do his utmost to take or destroy every vessel which it is his duty to encounter, or shall not do his utmost endeavor to afford relief to ships belonging to the United States, . . . shall, on conviction thereof by a general court martial, suffer death, or such other punishment as the said court shall adjudge.

Art. XIII. If any person in the navy shall make or attempt to make any mutinous assembly, he shall on conviction thereof by a court martial, suffer death; and if any person as aforesaid shall utter any seditious or mutinous words, or shall conceal or connive at any mutinous or seditions practices, or shall treat with contempt his superior, being in the execution of his office; or being witness to any mutiny or sedition, shall not do his utmost to suppress it, he shall be punished at the discretion of a court marital.

Art. XIV. No officer . . . in the navy shall disobey the lawful orders of his superior officer, or strike him, or draw, or offer to draw, or raise any weapon against him, while in the execution of the duties of his office, on pain of death, or such other punishment as a court marital shall inflict.

Art. XV. No person in the navy shall quarrel with any other person in the navy, nor use provoking or reproachful words, gestures, or menaces, on pain of such punishment as a court martial shall adjudge. . . .

Art. XVI. If any person in the navy shall desert to an enemy or rebel, he shall suffer death.

Art. XVII. If any person in the navy shall desert, or shall entice others to desert, he shall suffer death, or such other punishment as a court marital shall adjudge; and if any officer . . . shall receive or entertain any deserter from any other vessel of the navy, knowing him to be such, and shall not, with all convenient speed, give notice of such deserter . . . he shall on conviction thereof, be cashiered, or be punished at the discretion of a court martial. All offenses committed by persons belonging to the navy while on shore shall be punished in the same manner as if they had been committed at sea. . . .

Art. XIX. If any officer . . . shall, through intention, negligence, or any other fault, suffer any vessel of the navy to be stranded, or run upon rocks or shoals, or hazarded, he shall suffer such punishment as a court martial shall adjudge.

Art. XX. If any person in the navy shall sleep upon his watch, or negligently perform the duty assigned him, or leave his station before regularly relieved, he shall suffer death, or such other punishment as a court martial shall adjudge. . . .

In addition to the Articles of War and Navy Regulations (to be addressed subsequently), officers and men are governed by orders specific to their ship issued by their commander. The following are captain's orders issued aboard Constitution *directly pertaining to the character and conduct of an officer:*

1st—Strict attention must be observed to the printed instructions issued by the President of the United States.

10—The honors due the quarter deck, cannot be dispensed with on entering, either from below or a boat, and a polite address, and decent deportment from one officer to an-other is expected at all times thereon, for the character of a gentleman and an officer can never be separated.

22—Commissioned officers are requested to exact upon all occasions of duty from the warrant and petty officers, and they also from their inferiors, the most ready, unequivocal and respectable compliance with their orders, and it is expected of all inferiors that they do not neglect any exterior mark of respect whenever they address or are addressed by a superior upon duty.

27—The officers and petty officers are required to make themselves personally acquainted with the ship's company, in order to their being able to address them by their own names whenever they have occasion to call to them aloft or elsewhere.

58—The mates and midshipmen are ordered to sleep in hammocks, which are to be brought upon deck and taken down the same time [as] the ship's company.

59—Mates and midshipmen are to keep log books, a public order book, [a] clothes list of their subdivisions, Watch, Quarter and Station bills, clean and well written, all of which are to be brought to the captain the last Sunday in every month for his inspection.

65—Officers . . . of every denomination going out of the ship on duty are expected to be dressed in their uniform with swords.

85—No midshipmen is allowed to quit the deck at the expiration of the watch, under pretense that there is no one to relieve him, without having made such representation to the officer of the watch, and obtained his permission for so doing.

AN ADMONISHMENT TO MIDSHIPMEN

Edmund M. Blunt's recently published Seamanship in Both Theory and Practice *(1813) makes the following excellent observations about leadership by officers:*

Points of service ought to supersede every private consideration. The officer that does not feel a zeal and pride to excel in his department is unworthy of the rank he holds; for how can he expect more from the men whom he commands than he himself exhibits. A proper degree of spirit, mixed with encouragement, will always forward the duty better than oaths and coercion. It should therefore be the study of every . . . officer to instill with equanimity of temper, a spirit of emulation into his inferiors; which will in the event of success, or superiority, over any other ship of war, spread satisfaction over the whole; and all will in that case be as studious to excel as the officer to whose plans and arrangements the merit is attributed.

THE FRIGATE *CONSTITUTION*

The development of USS Constitution *and the vessels of her class is an interesting one with which you should be familiar as a naval officer. New to ships and the sea, some nautical terms may be unfamiliar to you. You may remedy this by consulting the glossary at the back of this volume.*

The United States Navy was disbanded following the Revolutionary War. However, with the new Constitution of 1789 and suffering the depredations of the Barbary pirates, Congress deemed it expedient to reconstitute a navy to defend U.S. interests at sea.

The Constitution of the United States of America, Article I:

Sec. 8: The Congress shall have the power . . .

13. To provide and maintain a navy;

14. To make rules for the government and regulation of the land and naval forces;

In 1794, the First Session of the Third Congress therefore approved:
An Act to Provide a Naval Armament.

Whereas the depredations committed by the Algerine corsairs on the commerce of the United States render it necessary that a naval force should be provided for its protection:

Section 1. *Be it therefore enacted by the Senate and House of Representatives of the United States of America in Congress assembled*, That the President of the United States be authorized to provide, by purchase or otherwise, equip and employ four ships to carry forty-four guns each, and two ships to carry thirty-six guns each.

Of the four 44-gun frigates originally authorized, three were built. The frigate United States *was launched at Philadelphia on May 10, 1797, the frigate* Constitution *at Boston on October 21, 1797, and the frigate* President *at New*

York on April 10, 1800. All three ships were designed by the Philadelphia shipwright Joshua Humphreys.

Realizing that his new nation had not sufficient resources to match the great European powers ship for ship, Humphreys designed his frigates to be longer and heavier than those concurrently being built in Europe. His idea, amply vindicated by the experiences of the present war to date, was to create frigates with sufficient hull strength and armament to overpower European frigates, but with sufficient speed to elude European line-of-battle ships.

CONSTRUCTION OF USS *CONSTITUTION*

President George Washington selected "Constitution" for the name of the frigate being constructed by Colonel George Claghorne at Edmund Hartt's Shipyard in Boston. The Boston naval agent, Henry Jackson, acquired the materials used to build the ship, which was constructed under the superintendence of her first commander, Captain Samuel Nicholson. Her construction budget was as follows:

	$
Timber, and every other material of wood, except masts	50,000.00
Labor, for building and fitting the hull, and rigging the ship	80,000.00
Smith's work including iron	21,000.00
Anchors	2,788.80
Masting	5,776.00
Sailmaker's bill	12,000.00
Carver's bill	800.00
Tanner's bill	500.00
Painter's bill	1,444.00
Cooper's bill	3,419.74
Blockmaker's bill	2,160.00
Boat Builder's bill	1,000.00
Cordage bill	37,000.00
Plumber's bill	1,444.00
Ship Chandlery bill	5,776.00
Turner's bill	577.60

Woolens for Sheathing	600.00
Making gun carriages	912.00
Cannon and Military Stores	28,880.00
Contingencies, Kentledge, Camboose, etc.	18,921.81
Sheathing, copper nails, and rudder braces	20,000.00
	[$294,999.95]

Construction of the frigate began in the summer of 1795. At the end of the year the secretary of war reported of Constitution'*s progress:*

The keel is completed and laid on the blocks. . . . The stern frame is now completing and will be soon ready to raise. The stem is also putting together. . . . About two thirds of the live oak timbers have been received, and [a] . . . great part of those timbers are bolted together in frames and are ready to put into the ship; but some of the principal pieces for the frame [*sic*] have not yet arrived. All the gun deck and lower deck beams are procured and are ready for delivery, and the plank[s] for those decks are received into the yard. The plank[s] for outside and ceiling are also received and are now seasoning. The copper is all in the public stores. The masts, bowsprit, yards and other spars, are all ready for working. The bits for the cables, coamings for the hatchways, partners for the masts are all ready. The caboose with a forge, hearth, armorer's tools, spare coppers, boilers, &c. are all complete; most of the iron work is in great forwardness; all the necessary contracts are entered into by the agent, and the articles contracted for are daily arriving.

By June 1797, two years into her construction, the secretary of war could report of a ship nearing completion:

The bottom of this ship is squared off, and the Caulkers are at work. The various decks are laying: the breasthooks, diagonal riders and counter timbers are all in and secured, and the mast makers are employed on the masts and yards. All the boats excepting the pinnace are built.

The riggers are at work on the rigging which will be soon ready: The water casks are in hand: Sails are preparing, and the constructor reports the ship may

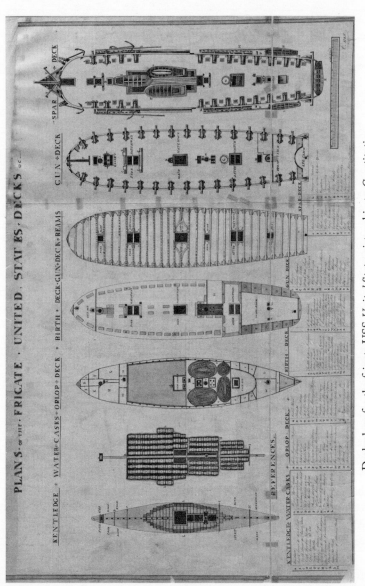

Deck plans for the frigate USS *United States*, sister ship to *Constitution*.

be launched about the twentieth of August next—the captain is of [the] opinion she may be completely equipped in one month after.

The estimates of both Colonel Claghorne and Captain Nicholson proved optimistic, however. Constitution *was launched on October 21, 1797, and entered service on July 22, 1798, the total cost of her construction being $302,719.*

USS *CONSTITUTION:* DIMENSIONS, CHARACTERISTICS, AND FEATURES

Constitution's *principal dimensions are as follows:*

Hull:
Length of hull overall, billet head to taffrail: 204 feet.
Length of hull, waterline, loaded: 175 feet.
Length of straight keel: 158.
Height of billet head above keel: 43.2.
Beam: 43 feet 6 inches.
Draft forward, loaded: 21 feet.
Draft aft, loaded: 23 feet.
Displacement: 2,200 tons.
Depth of hold: 14 feet, 3 inches.
Decks (lowest to highest): Orlop, Berth, Gun, Spar.
Height between spar and gun decks (deck to underside of beams): 6 feet, 1¼ inches.
Height between berth and gun decks (deck to underside of beams): 5 feet, 6½ inches.
Ballast: 140 tons.
Copper sheathed.

Masts and Sails:
Ship, length overall, flying jib to spanker boom: 306 feet.
Bowsprit, total length from billet head to end of flying jib boom: 76 feet.
Foremast, keel to truck: 198 feet.
Mainmast, keel to truck: 220 feet.
Mizzenmast, keel to truck: 172 feet, 6 inches.

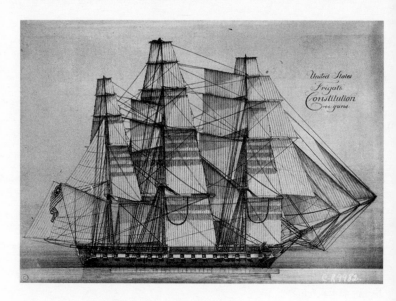

Foremast, height above waterline at load: 170 feet.

Mainmast, height above waterline at load, 186 feet.

Mizzenmast, height above waterline at load, 152 feet.

Foremast, greatest diameter, at spar deck: 29.5 inches.

Mainmast, greatest diameter, at spar deck: 32 inches.

Mizzenmast, greatest diameter, at spar deck: 23.5 inches.

Total number of sails, 33.

Sail area, 42,710 square feet (0.98 acres).

Maximum speed, 13 knots.

Armament:

Spar Deck: Two 24-pounder gunnades, as chase guns; twenty 32-pounder carronades.

Gun Deck: Thirty 24-pounder long guns.

Boats:

One 36-foot pinnace.

Two 30-foot cutters.

Two 28-foot whaleboats.
One 28-foot gig.
One 22-foot jolly boat.
One 14-foot punt.

Anchors:
Two main bower: 5,300 pounds.
One sheet anchor: 5,400 pounds.
Two kedge anchors: c. 500 pounds.

Complement:
Ship's Company: c. 450, including marines.
Provisions: 48,600 gallons of water; food for 6 months.

CONSTITUTION'S SIGNIFICANT CHARACTERISTICS:

Hull:
Constitution's decks from the lowest to highest are designed as follows:

The Orlop Deck, at the bottom of ship, is largely given over to storage. Forward it contains the sail rooms, store rooms, and the forward magazine, all in proximity to the foremast. Between the fore and main masts on the Orlop Deck lies the main hold. Next aft is storage for the anchor cables around the mainmast. Medical stores, the cockpit, and surgeon's mates' cabins lie between the main and mizzen masts. Storage for food and spirits occupies the area aft of the mizzen.

The Berth Deck, next above the Orlop, features the sick bay forward around the foremast. Aft of the foremast, and occupying more than half of the berth deck is the berthing area for able and ordinary seamen, with the marines and petty officers at the after end. Aft of the main mast is the officers' quarters, with cabins outboard starboard and larboard and the wardroom amidships. The sixteen midshipmen berth eight to a cabin in the starboard and larboard cabins located about half way between the main and mizzen masts just forward of the wardroom. These cabins measure nine feet athwart ships and eight feet fore and aft, with lockers lining the forward and after bulkheads.

The Gun Deck, next above the Berth Deck, aft to the mizzenmast, houses Constitution's *thirty long guns*. Along the centerline amidships lie the scullery, galley, *scuttlebutt, and grog tub forward of the main hatch with the large capstan aft of the main mast. The hawseholes for the anchor cables are at the extreme forward end of the Gun Deck, with the cables running aft to pass through the deck into the cable locker on the Orlop. Aft of the mizzenmast are the commodore's and captain's quarters.*

The Spar Deck, the uppermost, is largely devoted to serving the carronade guns, with the main hatch forward of the main mast, and the capstan aft of it. Upon the after end of the Spar Deck, called the quarterdeck, sits the ship's wheel and from this area the ship is commanded and maneuvered.

Masts and Rigging:

Constitution *is a three-masted, square-rigged vessel, the latter meaning that her sails are carried on yards suspended at close to right angles to the fore-and-aft midships line of her hull. Her masts from forward are the fore, main, and mizzen. Masts are supported and sails are carried on the yards and positioned to take best advantage of the wind by the use of rigging.*

Standing rigging is that which remains immobile and supports and stiffens the masts. It includes:

Stays lead forward (forestays) or aft (backstays) from the mast, either to the deck, another mast, or the bowsprit or jibboom. The forestays are also used to carry most of the ship's fore-and-aft staysails.

Shrouds brace a mast to starboard and larboard, running from the ship's sides to the masthead, with topmast shrouds extending from the top to the crosstree above. Small ropes called ratlines are tied across the shrouds to create the ladders that sailors use to climb aloft. Horse, also called footrope, is strung below yards for sailors to stand upon while adjusting sails.

Running rigging is lines rigged through pulleys, called blocks, used to do the work of lifting yards or loads and trimming sails. It includes:

Halyards, which are used to raise and lower yards.

Braces, which extend aft from yards and are used to trim them horizontally.

Sheets and tacks, which control the lower ends of sails.

Sails are named for the yards to which they are affixed. The lowest sail on the mast, the mainsail (or course), takes its name from the mast, viz. foresail or fore course and mainsail or main course. Next above are the topsails, followed by the topgallants, the royals, and finally the skysails. So: the fore course, the fore topsail, fore topgallant, fore royal, and fore skysail, and the same with the main sails and mizzen sails.

The lowest sail on the mizzen is the fore-and-aft spanker, used to improve maneuverability, as do the headsails set forward of the fore mast. Sails are generally set from lowest to highest and furled in reverse order, depending upon wind and sea conditions. Royals, skysails, and studding sails extended from the yards are all set to increase sail area in light winds. Staysails are often used during storms and are named for the stay upon which they are set. Sails are often reduced in battle, both to limit the damage to them and to reduce the number of men required aloft.

The above is, to be sure, an over-simple explanation. In truth, the best way for a midshipman or any sailor to "learn the ropes" is through practical application, starting with careful attention to the instruction offered by the lieutenants, sailing masters, and mates.

Constitution's *complement, provisioning, armament, and boats will be addressed individually in subsequent chapters.*

CONSTITUTION'S *NOTEWORTHY FEATURES:*

Constitution's *designer, Joshua Humphreys, incorporated into the American 44-gun frigate several elements that have proved exceedingly valuable in her service to date. These include:*

A longer hull with more length to beam than the two French-built British frigates, HMS Guerriere *and HMS* Java *that she defeated in 1812. That gives her a finer and (other conditions being near equivalence) somewhat faster hull than a European frigate and makes her appreciably faster than any European ship of the line.*

A hull constructed of white oak inner and outer planking, but with frames of live oak. Live oak is a much denser wood than other oaks, and is only available in the coastal forests of the American South.

Live-oak frames set two inches apart, rather than the customary four to eight inches used in European frigate construction. This makes Constitution's *hull less vulnerable to penetrating fire than her European counterparts.*

Diagonal riders rising from her keelson to support her gun deck. This additional support, rarely built into frigates, stiffens the hull to resist hogging, and transmits the weight of the guns atop the riders into the keelson amidships. As a consequence, Constitution *and her sisters carry 24-pounders on their gun decks, as opposed to the 18-pounders generally carried on the gun decks of European frigates. This grants* Constitution *a considerable advantage, both in weight of*

The diagonal riders supporting *Constitution's* gun deck.

broadside and in range, used to good effect by her crews in Constitution's *two victories to date.*

Taken together, her longer and finer hull, components of live oak, close-set frames, and diagonal riders that permit her to carry heavier armament, give Constitution *an extraordinary combination of speed, defensive strength, and offensive power, producing a rare vessel indeed: one stronger, more heavily armed, and potentially faster than the frigates of other navies she will confront in single-ship actions, but significantly faster than any hostile line-of-battle ships that she may encounter.*

MANNING *CONSTITUTION*

THE SHIP'S COMPANY

When originally commissioned, Constitution *was authorized by the Naval Act of 1797 to carry 364 officers and men. In 1807 the number was revised to 420. Although her manning varies slightly, depending upon the success of her recruiting, at the moment of publication* Constitution *carries 438 officers and men, viz.:*

Commissioned Officers: *One captain, one first lieutenant, four additional lieutenants, one chaplain, one surgeon, two surgeon's mates, one purser, two sailing masters.*

Commissioned officers are those gentlemen in the direct line of command, as well as others possessed of particular learnedness and character. The sailing masters navigate and trim the ship and keep her logs. The purser has charge of her stores and accounts.

Warrant Officers: *one boatswain, one gunner, one sailmaker, one carpenter, sixteen midshipmen.*

Warrant officers are master practitioners of particular manual skills essential to the operation and safety of the ship; they receive their warrants from the captain. Midshipmen, as apprentice officers, are also so classified.

Petty Officers: *three master's mates, two boatswain's mates, two carpenter's mates, one sailmaker's mate, fifteen quarter gunners, one captain's clerk, two yeomen, one coxswain, one armorer, one cooper, one master at arms, one cook, one steward.*

Petty (petit) officers assist their warrant officers to achieve their tasks. The yeomen assist with the accounting of supplies, the coxswain commands ship's boats in the absence of a commissioned or warrant officer, the armorer maintains and arms projectiles and small arms, the cooper makes and repairs barrels, and the master at arms is charged with maintaining good order and discipline.

Men: *218 Able Seamen, 92 Ordinary Seamen and twelve boys.*

Able seamen are those of the highest rating. Boys are youthful novices in naval service employed as servants, messengers, galley workers, and powder passers.

Marines: *one captain, one lieutenant, two sergeants, two corporals, one drummer, one fifer, forty-two privates.*

Constitution's *force of marines maintains order aboard ship, supports boarding parties, repels boarders, and joins landing parties.*

LIEUTENANTS

In this small volume, space will not permit detailed description of the functions of each man aboard. Since all midshipmen aspire to a lieutenancy, however, it may serve as an instructive example to enumerate the duties of same:

Of the duties of a Lieutenant:

He shall promptly, faithfully, and diligently execute all such orders as he shall receive from his commander, for the public service, nor absent himself from the ship without leave, on any pretense.

He is to keep a list of the officers and men on his watch, muster them, and report the names of the absentees. He is to see that good order be kept in his watch, that no fire or candle be burning, and that no tobacco be smoked between decks.

He is not to change the course of the ship at sea without the captain's directions, unless to prevent an immediate danger.

No boats are to come on board or go off without the lieutenant of the watch being acquainted with it.

He is to inform the captain of all irregularities, and to be upon deck in his watch, and prevent noise or confusion.

He is to see that the men be in their proper quarters in time of action; and that they perform all their duty.

The youngest lieutenant is frequently to exercise the seamen in the use of small-arms; and in the time of action he is to be chiefly with them.

He is to take great care of the small-arms, and see that they be kept clean, and in good condition for service, and that they be not lost or embezzled.

The first lieutenant is to make out a general alphabetical book of the ship's company, and proper watch, quarter and station bills, in case of fire, manning of ship, loosing and furling of sails, reefing of topsails at sea, working of ship, mooring and unmooring, &c., leaving room for unavoidable alterations. This is to be hung in some public part of the ship, for the inspection of every person concerned.

No lieutenant, or other officer, belonging to a ship of the United States, is to go on shore, or on board another vessel, without first obtaining permission from the captain or commanding officer, on his peril; and in the absence of the captain, the commanding officer is to grant no permission of this sort, without authority from the captain, previous to the captain's leaving the ship.

THE SHIP'S DIVISIONS

Mr. Blunt writes:

On Divisions

A ship's company, should be divided into three divisions, with their respective officers to each, viz. one Lieutenant, and two Midshipmen; and mustered at least twice a week, to see that their clothes are kept in good order and clean.

The usual method of forming the divisions, is by dividing the able seamen, ordinary, and landsmen, into three classes each, which makes an equal distribution of the men, as the same divisions form the boarders, sail-trimmers, and firemen in the quarter list; and the men by this plan are already selected for any sudden occasion, such as taking possession of a prize, landing, &c. as any one division, or part, can be immediately ordered on the service; on beating to divisions, the men should fall in, and be mustered and inspected by their respective officers, and reported to the 1st Lieutenant. . . .

One of Constitution's *captains ordered her divisions as follows:*

First Division—2nd Lieutenant, his division to consist of all the men stationed at the first division of guns on the lower Deck, and the fore topmen—

Second Division—3rd Lieutenant, his division to consist of the men stationed at the second division of guns on the lower deck, and the main topmen—

Third Division—4th Lieutenant, his division to consist of the men stationed at the third division of guns on the lower deck and at the mizzen topmen—

Fourth Division—1st Lieutenant, his division to consist of all the men stationed on the upper deck, together with those stationed at the magazine, shot lockers, pumps, galley and store rooms.

THE SHIP'S MARINES

Somewhat distinct from the ship's company is her force of marines, the duties of which are discussed below:

The lieutenant of marines is to exercise his men twice a day at sea when the weather permits; he is charged with placing all sentinels, and no sentry is to be relieved but by his order; he is required to pay strict attention to the cleanliness of the sentinels; he is to command the marines in time of action; he is to direct them in ceremonial honors in receiving company agreeable to the regulations directed in the "harbor rules." The marines in the routine of duty in a watch [are] under the direction of the officer of the watch in the same manner as any other part of the crew except in being sent aloft; the lieutenant of marines will pay particular attention in seeing his men's arms kept clean and in perfect order for action—the officer of the watch is directed to spare the marines for cleaning their arms when the duty of the ship will admit of it; but should any evolution for the ship require their immediate attention, the marines are directed to repair instantly to their station in the watch.

PAY AND PROVISIONS

Some sense of the expense of manning USS Constitution *may be gained from figures derived from a report to Congress issued by the Secretary of War in 1797:*

"An estimate of the pay and subsistence of the captain ... and crew ... of the Frigate ... Constitution of 44 guns. ... Pay of the officers, seamen and marines of [a] Frigate of 44 guns"

	$ Per Month	Twelve Months
1 Captain,	75	900
4 Lieutenants,	40	1,920
1 Lieutenant of Marines,	26	312
1 Chaplain,	40	480
1 Sailing Master,	40	480
1 Surgeon,	50	600
2 Surgeon's Mates	30	720

	$ Per Month	Twelve Months
1 Purser,	40	480
1 Boatswain,	14	168
1 Gunner,	14	168
1 Sail-maker,	14	168
1 Carpenter,	14	168
2 Boatswain's Mates,	13	312
2 Gunner's Mates,	13	312
1 Sail-maker's Mate,	13	156
8 Midshipmen,	13	1,248
2 Master's Mates,	13	312
1 Captain's Clerk,	13	156
1 Coxswain,	13	156
1 Yeoman of the Gun-room,	13	156
11 Quarter Gunners,	13	1,716
2 Carpenter's Mates,	13	312
1 Armorer,	13	156
1 Steward,	13	156
1 Cooper,	13	156
1 Master at Arms,	13	156
1 Cook,	13	156
150 Able Seamen,	11	19,800
103 Ordinary seamen and [boys],	9	11,124
1 Sergeant of Marines,	10	120
1 Corporal,	10	120
1 Drummer,	9	108
1 Fifer,	9	108
50 Marines,	9	5,400
For Frigate Constitution (359 crew).	4,080	$48,960

"Subsistence."

	Rations Per Day	Twelve Months
1 Captain,	6	2,190
4 Lieutenants,	3	4,380
1 Lieutenant of Marines,	2	730
1 Chaplain,	2	730
1 Sailing Master,	2	730
1 Surgeon,	2	730
2 Surgeon's Mates	2	1,460
1 Purser,	2	730
1 Boatswain,	2	730
1 Gunner,	2	730
1 Sail-maker,	2	730
1 Carpenter,	2	730
2 Boatswain's Mates,	1	730
2 Gunner's Mates,	1	730
1 Sail-maker's Mate,	1	365
8 Midshipmen,	1	2,920
2 Master's Mates,	1	730
1 Captain's Clerk,	1	365
1 Coxswain,	1	365
1 Yeoman of the Gun-room,	1	365
11 Quarter Gunners,	1	4,015
2 Carpenter's Mates,	1	730
1 Armorer,	1	365
1 Steward,	1	365
1 Cooper,	1	365
1 Master at Arms,	1	365
1 Cook,	1	365
150 Able Seamen,	1	54,750

	Rations Per Day	Twelve Months
103 Ordinary seamen and [boys],	1	37,595
1 Sergeant of Marines,	1	365
1 Corporal,	1	365
1 Drummer,	1	365
1 Fifer,	1	365
50 Marines,	1	18,250
For Frigate Constitution (359 crew).	383	139,795
139,795 rations, at 20 cents per.		$27,959

Pay and subsistence of the captain and crew of the Frigate *Constitution:* $76,919
War Office, June 16th 1797.   James McHenry. [Secretary of War]

Constitution's sailors are entitled to assign a portion of their earnings. From An Act for the Better Government of the Navy of the United States, *1800, Section 1, Art. XXXIV:*

Any person entitled to wages or prize money, may have the same paid to his assignee, provide the assignments be attested by the captain and purser; and in case of the assignment of wages, the power shall specify the precise time they commence. . . .

The requisite form for assigning wages or prize money appears below:

I _____ on board the United States vessel of war _____ commanded by _____ do by these presents allot _____ per month of my wages for the support of my family. And I do hereby appoint _____ my Attorney to receive for that purpose from the Navy Agent at the port of _____ the said sum of _____ monthly for the term of _____ months; the first payment to commence on the _____.

In witness whereof I have hereunto set my hand and Seal the _____ .

In presence and with the approbation of _____ Captain.

Registered by _____ Purser.

The daily provision of food and water aboard U.S. Navy ships has been prescribed as follows:

Sec. 8. *And be it further enacted*, That the ration shall consist of, as follows:

Sunday, one pound of bread, one pound and a half of beef, and half a pint of rice:—

Monday, one pound of bread, one pound of pork, half a pint of peas or beans, and four ounces of cheese:—

Tuesday, one pound of bread, one pound and a half of beef, and one pound of potatoes or turnips, and pudding:—

Wednesday, one pound of bread, two ounces of butter, or in lieu thereof, six ounces of molasses, four ounces of cheese, and a half a pint of rice:—

Thursday, one pound of bread, one pound of pork, and half a pint of peas or beans:—

Friday, one pound of bread, one pound of salt fish, two ounces of butter or one gill of oil, and one pound of potatoes:—

Saturday, one pound of bread one pound of pork, half a pint of peas or beans, and four ounces of cheese:—

And there shall also be allowed one half pint of distilled spirits per day, or, in lieu thereof, one quart of beer per day, to each ration.

SHIP'S DISCIPLINE

An Act for the Better Government of the Navy, *adopted June 1, 1800, also known as the Articles of War, lays down the rules of conduct for the officers and men of USS* Constitution, *particularly as concerning their conduct in battle. Other shipboard regulations are provided by the* Naval Regulations issued by Command of the President of the United States of America, *of 1802, and by orders issued by the captain to the crew of a particular ship. Elements of all of these sources of authority appear in this volume.*

The ship's master at arms is chiefly responsible for maintaining order, assisted by the boatswains and by the ship's marines. Breaches of good conduct are addressed in one of two ways. For lesser offenses captains may assign punishments including confinement or up to twelve lashes per individual offense. More serious transgressions shall be addressed by court martial, which may impose punishments ranging from stoppage of rations, wearing badges of dishonor, fines, loss of rank, confinement, suspension or dismissal from the service, floggings of up to one hundred lashes, to death.

Officers and midshipmen, being gentlemen, are exempt from corporal punishment, but may be admonished, confined, suspended from duty or cashiered, and are liable to suffer death for violation of the capital offenses enumerated in the Articles of War previously excerpted.

CAPTAIN'S ORDERS PERTINENT TO SHIP'S DISCIPLINE

7—Every superior officer is to inform and check those under him, of every impropriety he may see, or hear of their committing, either willfully or ignorantly, and to notify me of the same if it is of any magnitude.

23—Officers of every denomination are expected when the duty requires them at quarters on their different stations, to be particularly attentive to preserve silence among the men, and see that the orders issued from the quarter-deck are executed with celerity and without any noise and confusion.

24—It is my particular request that officers will upon all occasions, encourage and pointedly distinguish, those persons who are particularly cleanly, alert and obedient from those of different characters, in order that the deserving may see their merits are not disregarded, as well as the underserving be made sensible of the vigilance of the officers, and the advantages resulting from good and respectable conduct.

25—Discipline (as it respects alertness) can only be preserved by a constant and steady attention to whatever is carrying on; it is requested, the officers of the ship will not suffer the most trifling thing under them to be executed with indifference.

26—Blasphemy, profaneness, and all species of obscenity or immorality are peremptorily forbid and it is hoped the officers of every denomination will upon all occasions, discountenance and discourage such disorderly and despicable practices amongst the men.

57—Mates and midshipmen being acquainted with the established orders and regulations are expected zealously to enforce them and report any slackness or deviation they may observe in particular persons.

70—Upon the loss of any money, clothes, bedding, or other articles, the loser is immediately to make it known to the officer of the watch or commanding officer, who will take such measures as appear requisite to discover the thief. No man is permitted to appropriate to himself any clothes or other articles that he may at any time find about the ship, and if he cannot find the owner he is commanded to take it to the officer on the Quarter deck.

86—Each officer . . . is to keep a list of his division, and to be respectively responsible for the good order, cleanliness and sobriety of his division.

As for officers' use of punishments to maintain ship's discipline, Mr. Blunt offers the following wise observations:

Remarks on punishments

When people know that their actions are noticed, and that every reward or punishment is apportioned accordingly, they naturally become cautious of avoiding censure, and desirous to promote their own happiness and advantage.

Examples, at first should be severe . . . for by this a great deal of punishment will be obviated, which might otherwise become necessary. When rules and regulations, which beget order and discipline, are known and established, it requires very little trouble to keep them in force; the want of energy and vigilant attention are the only obstacles that would impede their operation. . . .

But the act of a moment should never betray an officer to the commission of injustice . . . as the distinction between guilt and error, should be the grand criterion of forming an officer's judgment on the propriety of punishing or acquitting the culprit. To confound the good with the bad would be to destroy all emulation, and every stimulus that would incite a man to exertions deserving of superior notice. When a prisoner is summoned, let each party be heard, and every circumstance fairly and impartially stated; if doubt exists, or it appears that the crime may have proceeded from ignorance, or mistake, the character of the accused should always be referred to, which will convince men of the importance of possessing a good name; but, if punishment must be inflicted, it should be done so as to strike others with a sense of its necessity, and the force of its example. . . . When a prisoner is released from his confinement, either in consequence of his innocence, or as an act of grace to his merit and former good character, it should be done as publicly as if punishment had been inflicted. Example, in all cases, is the grand object. . . .

To conclude this subject, [observation shows] that, where the discipline and subordination of a ship was regular, the people were happy; and the officers and ship's company, from an acquaintance with each other's system, acquired and maintained a reciprocal respect for, and confidence in each other.

MATTERS MEDICAL

INJURY AND DISEASE AT SEA

USS Constitution *rates a surgeon and two surgeon's mates to attend to sickness and injury among her crew. Routine matters of illness or injury are treated by the surgeon, surgeon's mates, or enlisted men known as loblolly boys, in the sick bay at the forward end of the Berth Deck.*

Sailors seeking treatments report for sick call at 9am. Those individuals judged sufficiently ill or injured to warrant bed rest will be confined to sick bay and their names placed on the daily binnacle list submitted to the captain. The binnacle list is so named because it is kept in a drawer on the ship's binnacle on the quarterdeck for ready reference by the officer of the watch, and none are exempted from duty save those on the list.

When Constitution *beats to quarters before battle, the surgeon, surgeon's mates, and loblolly boys assume their station in the cockpit on the Orlop Deck between the main and mizzen masts and prepare to receive casualties. As these occur, they are brought to the cockpit and treated in order of severity.*

Since ships smaller than frigates do not rate a surgeon, and since you may one day find yourself in command of such a vessel, some study of naval medical practices is in order. The following passages are excerpted from Plain Remarks on the Accidents and Diseases Which Occur Most Frequently At Sea:

Dysentery

After giving a dose of rhubarb, should it work off freely, give 30 drops of laudanum every 12 hours; should it not, give another dose of rhubarb, and then the laudanum, once in 12 hours, should the stools be frequent. As this disease is apt to occasion a great degree of weakness, the patient should be supported by a nourishing diet, and a wine glass full of decoction of bark, with 14 drops of laudanum, should be given every four hours. The following mixture is often very useful in obstinate dysenteries:—into a cup of vinegar put two tablespoonsful of common salt; stir it a few minutes, then pour off the vinegar, and put to it twice the quantity of hot water; let the patient take two table spoonsful of this mixture every three hours, as hot as he can sip it, until it works upon him freely.

Small Pox

In this disease, the patient should be kept as cool as possible, with light covering to the body, and if the weather be hot, bathe the skin frequently with cold water. Preserve a loose state of the bowels, by giving a dose of salts, or cream of tartar, or some other mild purgative, every other day. Abstain from animal food, spirituous liquors, and all stimulating substances, subsisting on rice and molasses, barley, flour, gruel, &c., and sweet or slightly acid fruits. This course is to be pursued as long as there is much fever, or until the pustules are filled and begin to turn yellow; the patient may then return to a nutritious diet, and take tonics, such as decoction of bark, Huxham's tincture, and elixir vitriol.

Measles

Abstain from animal food and spirituous liquors; adhere strictly to a low unseasoned diet; keep in a moderately cool atmosphere, and preserve a loose state of the bowels by taking castor oil or cream of tartar.

The Scurvy

Which is known by bad breath, loosened teeth, weariness, &c., requires warm clothing, wine, vegetables and fruit, such as oranges, lemons, tamarinds, apples, cocoa-nuts, onions, cabbages, and raw potatoes; malt tea, malt and spruce beer, cider, &c.; 15 or 20 drops of elixir vitriol may be taken now and then in a glass of wine and water, and the use of spirits must be avoided. A milk diet would be very good if attainable, and an abstinence from salted and smoke-dried provisions. Diseased gums are to be washed with elixir vitriol, so far diluted as to be agreeable to the taste, or with decoction of bark, or alum water. Costiveness should be avoided by taking a solution of cream of tartar frequently. To prevent the scurvy, particular care ought to be taken to avoid the use of fat or slushy food; for that reason, peas or beans may be boiled without allowing pork or other fat meat to be boiled with them. They may be salted so as to be palatable; and the like care ought to be taken of soups. Free exercise should be taken, by walking the decks, &c., in the open air, when there is no other employment to keep the men from below. Scorbutic ulcers may be dressed with raw potatoes soaked and made into a poultice with vinegar. . . .

Fractures

Should be treated pretty much in the same way as dislocations. When the bones are put in their proper place, carefully bathe the limb with brandy and vinegar, or brandy; then pass a flannel bandage a number of times round the limb, five inches above and five below the fracture; place upon the bandage four splints 10 or 12 inches long . . . at equal distance from each other, securing them with a string; bathe it frequently through the bandage with brandy, or sugar of lead water, and should it swell so as to make the bandage too tight, loosen it occasionally as may be necessary.

Cases That Most Strongly Require Bleeding

Are violent falls or bruises, especially when the head or breast is much affected; in pleurisy fever, likewise. Bleeding is oftentimes advantageous in other inflammatory fevers. As the loss of blood has a very debilitating effect, it ought to be avoided in all cases attended with much weakness. In bleeding, tie a garter moderately tight round the arm, two inches above the elbow; after the veins have filled, it is generally best to open the one that appears largest. The arteries, which are known by their pulsation, lie below some of the veins, and ought carefully to be avoided. If you feel carefully in the bend of the arm, rather nighest the under side, you can discover the beating and will endeavor, of course, not to open a vein directly over it. After bleeding, put a little lint on the opening, and bind it up with any soft bandage.

Wounds

In bad wounds, there is often a profuse bleeding, which requires the first and most particular attention. The application of dry lint and a bandage will often succeed, but if an artery of any considerable size is injured, and the blood spurts out largely, you must form some tight compression between the wound and the trunk of the body or heart, which will stop the discharge, till you can find the bleeding, and secure it with your needle. If it be an arm or leg that is wounded or taken off, take a strong handkerchief or large cord, and tie it moderately tight some way above the wound; if the injury is below the knee or elbow, it will be best to fix the cord two or three inches above the knee or elbow, and put a round short piece of wood beneath the cord; by turning this stick round, you tighten the cord till it stops the blood; let someone hold the stick in this position till you wipe the blood from the wound; then slack the cord till the blood spirts out, at which time fix your eye on the vessel; order the cord again tightened, and keep sight of the vessel, till you pass your crooked needle along side of the vessel, about a quarter of an inch deep and draw the thread half through; then enter the needle where it came out, and pass it up the other side of the blood-vessel, so that the point may come out near where you entered it; then draw the thread through, and tie it tight and it will stop the bleeding of the vessel; proceed in like manner for others; then take off the cord, and dress

A card from 1808 showing muscular anatomy. It was carried by Dr. Benjamin P. Kissam, U.S. Navy Surgeon.

© USS Constitution Museum Collection. U.S. Navy Loan

the wound with lint, bandage, &c. . . . If any wound bruise or, swelling grows painful, turns purple, or dark colored, and small blisters arise on or near it . . . then the is danger of a gangrene, to prevent which, let the part be scarified or pricked in several places with a lancet, and the scars dressed with basilicon. It may be wet with vinegar, brine, or sea salt. A cloth wet with vinegar, and some bark, No. 6, sprinkled in, may be put on it; also the bark must be given inwardly, as directed in fevers and agues…

Of Persons Apparently Drowned

When a person has remained more than twenty minutes under water, the prospect of his recovery is small; but we should not too soon resign the unhappy object to his fate, but try every method for his relief, as there are many well attested instances of the recovery of persons to life and health, who have been taken out of the water apparently dead, and remained so a considerable time, without showing any signs of life. In attempting to recover persons apparently drowned, the principal intention to be pursued is to restore natural warmth, upon which all the vital functions depend, and to excite these functions by the application of stimulants. First, strip him of his wet clothes, and dry him well; when he is dried, lay him between two or three hot blankets, and renew them as they grow cold. Rub him constantly with salt, warm ashes or coarse dry cloths, and rub on his wrists and ankles spirits of hartshorn, No. 32; and frequently apply the same to his nose. You may likewise apply bottles or bladders filled with hot water to his feet and armpits. While these external means of restoring heat to the body are going on, you must inflate the lungs as soon as possible with a pair of bellows . . . or for want of bellows, you may use a common glister-pipe, or in case of necessity, a common tobacco pipe or quill. . . . When the lungs are full, press upon the breast, and force the air out again, and then blow as before. Repeat this process for half an hour or more. In addition to this method, you may dip a blanket into boiling water, wring it as dry as possible, and wrap the person in it. Repeat this every 15 or 20 minutes for two hours or more. These means ought to be continued for two or three hours at least, even if no signs of life appear. When signs of returning life are apparent, the frictions must be continued, but more gently; when the patient can swallow, he must take some warm spirits; when he is pretty well recovered, put him into bed in blankets, and give some warm spirits; if his feet should be cold, wrap them up in warm flannels.

CLEANLINESS AT SEA

Cleanliness—Wet and Damp Clothes

In order to prevent fevers, scurvies, and in a word a great part of the diseases that sailors are subject to, particular attention ought to be paid to cleanliness, and to avoid wearing or lying down with damp or wet clothes on, and also the intemperate use of spirituous liquors. Masters ought to have strict regard to their sailors on passing from temperate into tropical climates, especially on approaching the land, by not permitting the men, in watering parties, &c., to keep on shore without safe shelter, as they are thereby much exposed to the fall of heavy dews, which endanger their health. It is a good practice to evacuate well by purgatives on such a change; and if the men are in very high health, blood-letting has been practiced with the happiest result.

Another important element in ensuring health aboard ship is the cleanliness of the vessel herself:

Particular Points Requisite to Keep a Ship Clean

Whatever conduces to the preservation of health on board a ship should be attended to with the utmost strictness. . . . Excessive washing, in improper season and climates, will produce damps which occasion colds, the forerunner of most disorders generated on board ship; but no washing at all must render a ship loathsome, by suffering the generation of filth and vermin.

The lower deck of every ship, but more especially that of frigates and sloops of war, ought to be particularly attended to; as the small space which contains so many men requires the greatest attention for preserving it clean and well aired. . . . A free circulation of air is the most essential requisite for the preservation of salubrity below. . . .

CAPTAIN'S ORDERS PERTAINING TO THE HEALTH OF THE CREW

39—The first lieutenant or commanding officer, whether at sea or in harbour, is to visit the ship throughout every forenoon, to see that the tiers, cockpits, wings, store-rooms, passages & C&C are clear, clean and in proper condition, and report is to be made to the captain when ready for his visitation.

40—The between decks are to be cleaned every forenoon, to be washed twice a week, and fires shall be made, over which sentinels are to be placed and the people not permitted below until the decks are dry; the cockpit is to be cleaned whenever the between decks are, as are all ladders, gratings, port sills, coamings, hatchways, store-rooms & magazines, passages & wings.

42—Every day after dinner and just before the hammocks are piped down in the afternoon, the decks are to be swept.

48—The main deck is to be constantly kept as clear as possible and no chests, empty casks, or lumber of any kind suffered to remain on it—a great object being the health of the men. Dry air should be admitted especially in Summer or warm climates, but rain and the wash of the Sea should be avoided, and air that is damp admitted with great caution, a thorough draft of air in the channel is not necessary for any length of time, and should be introduced in preference at such times as the men are employed on deck.

49—The ventilation is to be constantly worked by day in winter, and by night and by day in summer or a warm climate, the wind sails should be used at all times in the summer when the air is dry and at times in the winter when the weather is fine.

55—Any person finding himself ill is to make his complaint without loss of time to the Surgeon's mates, as no excuse for neglect of duty on the score of illness will be received but through the Surgeon.

56—Washing days will be appointed as the weather and duty of the ship will admit and then things are to be dried on lines stretched from shroud to shroud forward, or in such other places as the commanding officer shall direct.

71—The Officers of the different divisions will order the petty officers under their command occasionally to examine into the state of the people's mess utensils of their division, to see that they have a sufficiency of those Articles and that they are kept wholesome and clean.

Constitution's *surgeon also notes the fumigation in ship's spaces with muriatic acid gas, vinegar wash downs, and the application of whitewash, all to promote and maintain cleanliness.*

Beyond questions of cleanliness, it is important to remember that danger aboard ship is both constant and ubiquitous, as the following brief illustrations attest. One of Constitution's *log entries of earlier years reads:*

Departed this life, John Hornock Boatswain, occasioned by being shot with a pistol that went off accidentally. . . . Committed the body of the deceased to the deep with the usual ceremony as performed at sea.

Death claims even an experienced man in a careless moment. The ship's surgeon recorded another misadventure just south of Newfoundland:

At 3 P.M. a sailor fell overboard out of the main chains. The topsail was instantly backed and the stern boat lowered down. The man being (fortunately) an expert swimmer, kept on top of the water, and was picked up about 200 yards astern. He said he could have taken off his shoes, but did not wish to lose them! The blood however appeared to have forsaken his cheeks. The tenure of a sailor's existence is certainly more precarious than any other man's, a soldier's not excepted.

This sailor was most fortunate. You would do well to heed the surgeon's words. As always, whether handling weapons or working aloft or over the side or engaged in any shipboard routine, be upon your guard!

PROVISIONING *CONSTITUTION*

REGULATIONS TO BE OBSERVED RESPECTING PROVISIONS

1. Provisions and slops are to be furnished upon the requisitions of the commanding officer founded upon the purser's indents.

2. The pursers being held responsible for the expenditures, shall, as far as may be practicable, examine and inspect all provisions offered to the vessel, and none shall be received that are objected to by him, unless they are examined and approved of by at least two commissioned officers of the vessel.

3. In all cases where it may appear to the purser, that provisions are damaged or spoiling, it will be his duty to apply to the commanding officer, who will direct a survey, by three officers, one of whom, at least, to be commissioned.

4. If upon a settlement of the purser's provision account, there shall appear a loss or deficiency of more than seven and a half percent upon the amount of provisions received, he will be charged with and held responsible for such loss or deficiency exceeding the seven and a half percent, unless he shows by regular surveys that the loss has been unavoidably sustained by damage or otherwise.

5. Captains may shorten the daily allowance of provisions, when necessity shall require it, taking due care that each man has credit for his deficiency, that he may be paid for the same.

6. No officer is to have whole allowance while the company is at short.

7. Beef for the use of the navy is to be cut into 10 pound pieces, pork into 3 pound; and every cask to have the contents thereof marked on the head, and the person's name by whom the same was furnished.

8. If there be a want of pork, the captain may order beef, in the proportion established, to be given out in lieu thereof, and vice versa.

9. One half gallon of water [per day] at least shall be allowed every man in foreign voyages, and such further quantity as shall be thought necessary

on the home station, but on particular occasions the captain may shorten this allowance.

10. To prevent the buying of casks abroad, no casks are to be shipped which will want to be replaced by new ones before the vessel's return to the United States.

11. If any provisions slip out of the slings, or are damaged through carelessness, the captain is to charge the value against the wages of the offender.

12. Every ship to be provided with a seine, and the crew supplied with fresh provisions as it can conveniently be done.

The scale of inventory necessary to subsist the crew of Constitution *may be understood by examining a list of only some of the provisions carried by a frigate:*

Packages, No. Quality	Quantity	Nature of the Provisions &c.
33 Barrels, 20 Tierces	12,600 Pounds	of Salt Beef
65 Barrels	13,000 Pounds	of Salt Pork
69 Hogsheads, 35 Bags	———Pounds	of Bread
8 Bags, 3 Firkins	622½ Pounds	Butter
15 Barrels	2,940 Pounds	Flour
6 Tierces	———Pounds	Rice
———	2,240 Pounds	Salt Fish
31 Casks	3,580 Pounds	Cheese
4 Casks, 1 Barrel, 5 Bags	53 Bushels	Beans
1 Barrel, 6 Bags	23½ Bushels	Peas
	203½ Bushels	Potatoes
———		
10 Hogsheads, 9 Tierces, 6 Barrels, 6 Half Barrels	1801½ Gallons	Rum
1 Hogshead	117 Gallons	Molasses
12 Barrels	408 Gallons	Vinegar
———	20 Bushels	Salt
12 Boxes	590½ Pounds	Tallow Candles
6 Boxes	315 Pounds	Soap
———	4 Chaldron	Coals
———	12 Cords	Wood
100 Butts, 20 Hogsheads, 30 Gang Casks	16,400 Gallons	Water

⚙

The ship's water may be resupplied by collecting rainwater while at sea or by dispatching landing parties to refill the ship's water casks at ports of call. Fresh produce and livestock are acquired before sailing and at ports of call during a cruise. Livestock may be kept alive aboard ship for some weeks into a voyage, and the ship's diet may also be supplemented with fish caught with the seine provided.

As evidenced by the above inventory, the crew is also entitled to a grog ration of one pint daily. This is served in half-pint portions after dinner and supper and consists of either rum or whiskey mixed with an equal portion of water.

REGARDING SLOPS

Slops are clothing and bedding issued to sailors from ship's stores. As per the Naval Regulations of 1814:

Regulations Respecting Slops

Seamen, destitute of necessaries, may be supplied with slops by an order from the captain, after the vessel has commenced her voyage....

Slops are to be issued out publicly and in the presence of an officer, who is to be appointed by the captain, to see the articles delivered to the seamen and others, and the receipts given for the same, which he is also to certify....

The captain is to oblige those who are ragged, or want bedding, to receive such necessaries as they stand in need of.

Captain's Rules Regarding Clothing and Bedding:

The Officer of each division is on the 27th day of each month to make a return to the Captain of the necessary clothing wanted by the men of his division in order that it may be issued on the last day of the month; as clothing will not be delivered excepting on those days.

19—All sorts of slops and clothing must be issued only on a Special Order from me.

76—The ship's company are to be mustered every evening at sunset ... and proper notice [is] to be taken at such times of men that are dirty or slovenly ...

and clothes however old should not be ragged. . . . The last Thursday and Sunday of every month [are] appointed as days of general review or muster of clothes, when the officers are expected to examine very particularly into the people's clothing and bedding and are to report to the captain any deficiency that may appear.

Captain's Orders for the Dress of the Crew:
Every seaman belonging to the ship is expected to supply himself with the following list of clothes:

Jackets, 2
Waist coats or inside jackets, 2
Pairs of trousers, 2 blue, 2 white
Shirts, 4
Pairs of stockings, 2
Black silk handk'fs or neck cloths, 2
Hats, 1 hat or 1 hat & 1 cap
Pairs of gloves, 2
Pairs of drawers, 2

In addition to the list of clothes above named the ship's company are expected to supply themselves with white outside jacket and waist coat to be worn occasionally in summer or a warm climate, with long white trousers.

SHIP'S STORES
Food and clothing are only one of the requirements of a frigate, which must also carry the supplies necessary to accomplish her mission. Below is an inventory of the tools necessary to maintain the ship's casks and barrels:

Cooper's Tools &c.
2 Adzes
2 Drawing Knifes
2 Hand Shaves

2 Hammers
2 Iron Drivers
2 Bung Borers
2 Tap Do.
2 pair Compasses
1 Bitt Stock
2 Cold Chizzels
4 Punches
2 Crows
1 Stock for do.
1 Flagging Iron
2 Brass Cocks
1 Han[d]saw
2 Do. Files
6 Dowel Bitts
1 Jointer with 2 Irons
1 Coopers Ax
1 Bick Iron
450 Rivets
2 Bundles Iron Hoops
1 pr marking Irons
1 Vice
12 lb Chalk

And the cooper's 26 separate categories of items are as nothing compared to the 130 categories listed in the "Inventory of Stores, Tools and other small articles in the Carpenter's Care."

In sum, Constitution *presents a continuous supply problem, from sails and stores and shot and slops—to stationery.*

❁

The following Articles of Stationery are required for the use of the Commander of U.S. frigate Constitution—viz.

One Ream Foolscap Paper
Half Ream Letter Paper
One hundred Quills

Boston, October 24, 1814

Approved Chs. Stewart
Approved Wm Bainbridge

SAILING *CONSTITUTION*

SHIPBOARD ROUTINE WHILST AT SEA
Aboard ship the twenty-four hours of the day are divided into watches, thus:

Morning: 4 to 8am.

Forenoon: 8am to noon.

Afternoon: noon to 4pm.

First Dog: 4 to 6pm.

Second Dog: 6 to 8pm.

Evening: 8pm to midnight.

Middle: midnight to 4am.

The dog watches, set around the evening meal, are used to stagger the larboard and starboard watches day by day. The crew are divided into alternate watches, starboard and larboard, and various specialists who do not stand regular watches (e.g., the surgeon, sailmaker, cook, etc.), known as idlers. Watch standers are either on watch or watch below, subject to summons at any hour the working of the ship requires. Each man's place and role for various evolutions are kept on the Watch, Quarter, and Station Bill, as illustrated by Mr. Blunt below.

Form of a Watch Bill, to be Hung on the Main Deck for the Starboard Watch

The same form for the Larboard Watch, is to correspond exactly in numbers, with a column for the number of the hammocks.... The number of the hammock may represent the person in the station bill, which will make it PERPETUAL, and prevent a great deal of trouble in erasing names on every change. This may be exemplified, by supposing a boatswain's mate, belonging to the starboard-watch whose number is one; and, on a large sheet of paper, the table, as follows,

No. Hammocks.	Quarters [battle].	Mooring & Unmooring.	Reefing & Furling.	Making & Short. Sail.	Tacking & Wearing.	No. of Division.
1	Main Rigging.	Starb. side Main Deck.	Starb. side Main Deck.	Starb. side Main Deck.	Starb. side Main Deck.	First

may be constructed; with every man's station opposite to his number in the marginal column.

MANEUVERING THE SHIP

Ships are steered by compass points. The 360 degrees of a compass are divided into 32 points, meaning an individual point equals 11.25 degrees. The compass points working from north to east, for instance, are as follows:

1. *North*
2. *North by east*
3. *North-northeast*
4. *Northeast by north*
5. *Northeast*
6. *Northeast by east*
7. *East-northeast*
8. *East by north*

followed by East, and so forth, as appear in the illustration on p.62:

Since a square-rigged ship can sail no closer than six points (68 degrees) to either side of the direction of the wind, to make any progress in an upwind direction she must sail somewhat perpendicular to the wind, altering course to and fro across the wind to reach her destination. This is called tacking; the direction from which the wind is coming indicates whether a ship is on a starboard or a larboard tack.

To maneuver the ship from a larboard onto a starboard tack it will be necessary to either put the ship's bow through the wind, tacking, or to temporarily reverse course and put the stern through the wind on a longer turn round to the other tack, called veering or wearing. Tacking has the advantage of expedition, but, since the ship will be taken aback, is harder on the rigging and may fail if not executed with sufficient dexterity.

Six divisions of sailors are employed to handle sail: one each in the fore, main, and mizzen tops, and three on the Spar Deck, the forecastle, waist, and afterguard.

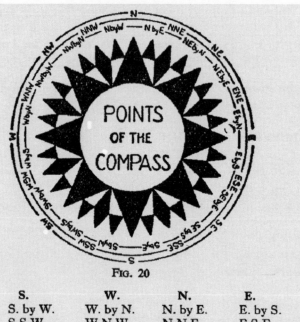

FIG. 20

S.	W.	N.	E.
S. by W.	W. by N.	N. by E.	E. by S.
S.S.W.	W.N.W.	N.N.E.	E.S.E.
S.W. by S.	N.W. by W.	N.E. by N.	S.E. by E.
S.W.	N.W.	N.E.	S.E.
S.W. by W.	N.W. by N.	N.E. by E.	S.E. by S.
W.S.W.	N.N.W.	E.N.E.	S.S.E.
W. by S.	N. by W.	E. by N.	S. by E.
W.	N.	E.	S.

The most active and skilled sailors are employed in the fore and main tops to unfurl and furl sails, the least so in the waist and afterguard to do the heavy hauling of running rigging.

Below are reproduced Mr. Blunt's instructions for tacking and veering ship, the two most consequential maneuvers at sea. His descriptions are not presented here with the idea that you will thereby learn everything needful, a thing best accomplished through observation and participation. Rather, they are offered to give you a sense of the complexity of the undertaking:

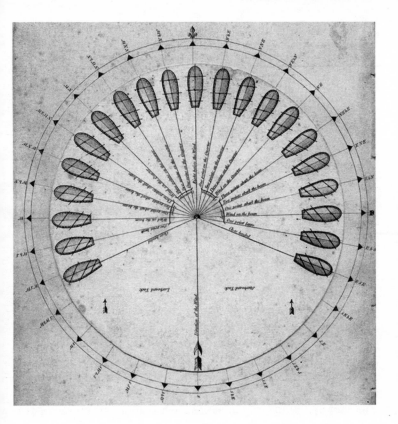

To Tack a Ship in Getting to Windward as Much as Possible.

To execute this with propriety, care must be taken that the ship does not yaw, that she is not too near or too far from the wind; because both situations are equally prejudicial.

When this medium is obtained, haul the mizzen out, while you put at the same time the helm a-lee, and brace the bow-line quite to leeward, that the mizzen may be as much as possible exposed to the wind. When the ship is come to the wind, so as to cause the square-sails to shiver, let go the jib and all the staysail sheets before the main-mast: at the moment when all the sails catch a-back, and particularly the mizzen-topsail, let it be braced sharp about the other way; hauling up at the same time the weather-clue of the main-sail, and trim sharp for the other tack as fast as possible. The jib and staysail sheets are also to be shifted over at the same time, in righting the helm, whether the ship has lost her way, or even still advances a-head. Then, as soon as she has passed the direction of the wind about 45°, in continuing her evolution, shift the foremast's sails, which are to be trimmed with the same celerity as in putting the helm a-lee, if you fear the ship (which must still go a-stern if the operation be slowly executed) will not fall off sufficiently: for, if the sails are braced about briskly, she will never have sternway; on the contrary, she will get a great deal to windward.

To Veer a Ship Without Losing the Wind Out of her Sails.

To execute this evolution both the main-sail and mizzen must be hauled up, the helm put a-weather, and the mizzen top-sail a shivering, which will be kept so till the wind be right aft, suppressing for that purpose the effect of all the staysails abaft the centre of gravity. As the ship falls off, (which she will do very rapidly,) round-in the weather-braces of the sails on the fore and main mast, keeping them exactly trimmed to the direction of the wind, and remembering also that the bowlines are not to be started till the ship begins to veer. As she falls off, ease away the fore-sheet, raise the fore-tack, and get aft the weather-sheet, as the lee one is eased off; so that, when the ship is right before the wind, the yards will be exactly square. Then shift over the jib and staysail sheets; and, the ship continuing her evolution, haul on board the fore and main tacks, and trim all sharp fore and aft, remembering to haul aft the mizzen and mizzen-staysail sheets as soon as they will take the right away, or when the ship's stern has a little passed the direction of the wind. When the wind is on the beam, right the helm to moderate the great velocity with which the ship comes-to; the sails being trimmed, stand on by the wind.

Maneuvering a sailing ship is the work of many hands. For furling the sails of a 44-gun frigate, Mr. Blunt gives the numbers of men as follows:

Main yard: 40.
Main top sail yard: 32.
Main top gallant yard: 6.

Fore yard: 38.
Fore top sail yard: 30.
Fore top gallant yard: 6.
Mizzen top sail yard: 18.
Mizzen top gallant yard: 6.
To furl the spritsail: 14.
To stow the jib: 9.
To stow the mizzen stay sail: 4.
Total number of men stationed: 203.

The orders issued by one captain of Constitution *show the disposition of midshipmen during sailing maneuvers:*

Whenever all hands are called for the purpose of making or taking in sail, tacking or wearing Ship &c....

Mr. Reed on the fore-castle
Mr. Deacon, fore and foretopsail Braces Starboard Side

Mr. Morris, fore and foretopsail Braces Larboard Side

Mr. Izard near the commanding Officer

Mr. Haswell near the commanding Officer

Mr. Dexter to attend Signals

Mr. Burrows to attend, Main and Maintop sail Braces Starboard side

Mr. Baldwin to attend, Main and Maintop sail Braces Larboard Side

Mr. Hunnewell, Cross Jack braces Starboard Side

Mr. Bartell, Cross Jack braces Larboard Side

Mr. Nicholson Fore top—

Mr. Rowe Main top—

Mr. Thompson Mizzen top— . . . .

Mr. Hunt Gun deck Fore sheet, and main tack Starboard side

Mr. Laws Gun deck Fore sheet, and main tack Larboard Side

Mr. Davis Main sheets Starboard side

Mr. Hall Main sheets Larboard side

CAPTAIN'S ORDERS PERTAINING TO SAILING AND SEA ROUTINE

2nd—The time by glass, must be regularly attended to, by night and by day; and the bell, at the call of the quarter master from the lee gangway, must be struck every half hour, and rung every four hours.

3rd—The log is to be hove at sea every hour, and the rate of going, courses, winds and occurrences instantly marked on the board, and at noon each day, the same are to be entered on the log-book by the master.

4—A good look out to be kept, night and day, and the officers are always to cause to be kept clear, the cannon, and ropes of every description, and to have every man at his Station.

5—The pumps are to be sounded by the carpenter or one of his mates every two hours, or oftener, if necessary, and the ship pumped out, whenever there are three inches more water in the well than the ship sucks at.

6—On seeing a vessel or vessels of any description whatsoever, as also on a change of the wind by night or by day, I must be immediately informed thereof, in order to determine on the line of conduct I may judge proper.

8—The officer of the watch will be ever attentive in seeing the sails neatly trimmed, and that no ropes are towing overboard; at day light in the morning, the stock are to be fed, and every evening before sun set, previous to the decks being swept and wetting the ship inside and out, which is to be done every morning & evening.

9—A sea lieutenant is always to be on the quarter-deck, together with a midshipman and quarter-master.

11—The weather side of the quarter-deck is reserved for the walk of the captain, or in his absence, the commanding Officer of the watch.

12—The time glass is to be regulated every evening by the watch; and the quarter-deck is always to be kept clear of clothes, lumber and dirt, the ropes flemished, coiled and hung up on the pins.

21—The sails are always to be neatly handled, in the order directed at the time, the yards squared, stay, back stays, and every rope hauled taught, and the decks and sides kept clear of dirt.

35—The mates and midshipmen are to be at three watches at sea and in port, one third of which may be absent on leave at a time.

72—Boats' crews are commanded to obey with as much alacrity and punctuality the orders of the coxswains as those of any other officer on board, no excuse will therefore be received for a boat being left on shore, or for anything lost or damaged belonging to a boat unless previously reported to the commanding officer.

100—In tacking, wearing or reefing the topsails, getting up or down top gallant masts or yards in the daytime the first lieutenant or day officer is to take the direction on the quarter deck altho' all hands are not called, and the officer of the watch to take such station as is necessary to the ready execution of the duty to be done, but the officer of the watch will not understand this as precluding him from tacking, wearing or reefing topsails or getting down top gallant masts or yards, when the occasion requires immediate execution.

106—In going into ports or coming out of ports, all Lieutenants & midshipmen are requested to be on deck in their respective stations to preserve silence and carry the Orders of evolution in execution with celerity and without confusion. . . .

OF SIGNALING

Communicating with other ships at sea being essential, a system of signals has been developed, as follows:

On Day Signals.

All signals, to be effectual, must be simple, and composed in such a manner as to express the same significance at whatever mast-head, or yard-arm, they may be displayed from. The following day signals will be found to have these advantages.

The plan is, to express numbers by distinctly coloured flags, each number referring to a certain signification, to be agreed on before-hand.

Mode of expressing 999 Numbers by Eleven Flags and One Pennant.

. . . . There are ten flags, each flag representing the number placed against it, and a substitute flag representing the same number with any flag hoisted next above it. To express from 9 to 99, hoist the flag standing for the first figure of the given number, above the flag standing for the second; that is, to express 45, hoist flag 4, above flag 5;—but should the given number be two similar figures, for instance 55, it is to be expressed by hoisting flag 5, above the substitute. To express from 99 to 999, hoist the flags one above the other in the order of the figures of the given number;—thus, 245 is expressed by hoisting flag 2, above 4, above 5; and 225 by flag 2, above the substitute, above 5; and 522 by flag 5, above 2, above the substitute. But as there are some instances in which the eleven flags are insufficient to express numbers above 99, a short thick pennant, denoting that the last figure of the given number is the same as the first, is proposed to remedy the defect; therefore, to express 545, which is a number that could not be expressed by the eleven flags, hoist flag 5, above 4, above the pennant; also to express 444, hoist flag 4, above the substitute, above the pennant.

This code includes a total of 247 day signals, the following score of which should suffice to illustrate their variety and applications.

Table of Significations.

0. A general acknowledgment that the signal made is understood.

3. Veer, sternmost and leewardmost first.

4. Make sail, if in close order headmost first.

7. Ships astern make more sail.

13. Prepare to anchor.

23. Carry all the sail you can without risk.

29. Disperse, and each vessel do the best for her safety.

75. Compass signification West by North.

86. Request the assistance of a surgeon.

97. We are in want of provisions.

108. We have sprung a lower mast, yard, or bowsprit.

129. Keep your station.

133. You mistake my signal.

144. Proceed on the service previously made known to you.

146. How many days' water have you on board?

152. Do you gain upon the chase?

170. The chase is a frigate.

179. Prepare to sail.

191. Engage generally.

216. Tack in succession, in the wake of your next ahead.

On Night Signals.

Night signals should be used as little as possible, since they are frequently misunderstood. Of necessity, they must be composed either of sound or light, or the two blended together. If several lights are shown together, that they may have the same appearance from every horizontal situation, it will be necessary to hoist them in a vertical position. In the following system of night signals, this circumstance is attended to: the plan is to express numbers by means of guns, lights, and blue lights. A single gun being left at liberty to draw the attention of the fleet, and to point out the position of the commodore.

OF FIRE AT SEA

Truer words were never spoken than those offered by Constitution's *surgeon:*

"The cry of fire is dreadful on *shore*, but ten thousand times more distressing on board a powder ship *at sea*."

Both preparation and vigilance are required to forestall such calamity:

Remarks on the Necessity of a Fire Bill, &c.

We now come to the most important of all the regulations of a ship, namely, those which operate against the fatal and shocking effects of fire . . . where in many cases, no alternative exists between the fury of the flames and that of the ocean, which very naturally excites at the moment in which it bursts forth, the most violent agitation: and . . . has frequently plunged many unhappy victims into a watery grave, at a time when, by a proper degree of coolness and caution, their lives might have been preserved, and the flames overcome.

To reflect on the event, and consult the best means of providing against it, appear to be the most likely means of preparing the minds of men to receive the shock of an alarm undaunted. This, admitted, will show the propriety and necessity of having a FIREBILL: for, whatever good results from stationing people in ordinary cases, cannot be put in competition with this, which provides against the most dreadful catastrophe incident to a ship. . . .

Proportions for stationing the ship's company in cases of fire . . . Class of ships, 44

At the Cistern Pump: 6.

To man the Runners or stays at the Fore Hatchway: 26.

To man the Runners or stays at the Main Hatchway: 26.

To Draw Water on the Starboard Gangway: 14.

To Draw Water on the Larboard Gangway: 14.

To hand Water from the Cistern Pump: 16.

To hand Water from the Starboard Gangway: 16.

To hand Water from the Larboard Gangway: 16.

To hand Water from the Hatchways to below: 22.

In the storerooms and passages: 11.

To work the engine: 6.

Party with the master: 20.
Party with the Boatswain: 15.
Party with the Carpenter: 5.

Explanation of the Fire Bill.

The forecastle men and quarter masters are to be divided under the command of the master and boatswain. The former is, with his party, to be employed in removing any stores, &c. to which the fire may be likely to communicate. The latter is, with his party, to collect wet hammocks, sails, &c. to smooth it. The gunner, with his crew, to be employed in the store-room, in removing stores and extinguishing the fire there. The top men, in . . . frigates, on the gangways, drawing water (which should be emptied into tubs, for the purpose of being handed from thence) at the cistern pump, and to work the engine. The remainder to assist the after guard, in passing it along, and returning the empty buckets. The waisters, and some part of the after guard, at the fore and main hatchways, to hoist anything up from below which may be necessary to remove. The Carpenter, and all his crew, to attend with their tools, to be ready to perform any service which the exigency of the case may require, and the marines should be divided round the ship with loaded musquets and fixed bayonets, to prevent any persons from leaping overboard.

N. B. Care should be taken, upon the discovery of fire, to prevent any air getting to it. The ports should be lowered, and the gratings laid on with tarpaulins over them; oatmeal has been found a very good thing to smother fire, as well as wet swabs, hammocks, and sails.

BOATS AND BOATMANSHIP

Your first independent command likely will be of a small boat. Ship's boats are used to ferry persons and supplies to and from shore, communicate between ships while at sea, and may be used to recover a man overboard. Constitution *at present carries eight boats, as follows:*

1, 36' Pinnace (a long boat)
2, 28' Whaleboats
1, 28' Gig
1, 22' Jolly Boat

1, 30' Cutter
1, 30' or 28' Cutter
1, 14' Punt,

located as follows—

1 Pinnace on Main Spar Deck hatch [for the use of landing parties, obtaining wood and water at ports of call and at islands visited].

2 Cutters nested in Pinnace, or one Cutter nested in Pinnace and 1 Cutter located in chocks on Main Spar Deck hatch beside Pinnace.

1 Gig, stowed at stern wooden davits.

2 Whaleboats located in wooden davits, [on the starboard and larboard] quarters.

1 Punt carried abaft of smoke pipe hatch for side cleaning and painting.

[1] Jolly boat . . . carried on an outrigger to [the] stern davits.

When ordered out in command of a boat by the officer of the deck, you will need to know the following commands for raising and lowering a boat and for making way once afloat:

To lower a boat:

"*Stand by your lines*": *crew on deck prepares to lower the boat using the lines attaching the boat to the davits, called falls.*

"*Slack away together*": *crew on deck eases off on the lines, letting the weight of the boat lower it and being careful to keep it even and to check the speed of its descent.*

"*Avast*": *crew on deck ceases to let the falls run and prepares to secure them once the boat is released from its falls and away.*

To raise the boat once reattached to its falls:

"*Haul away together*": *crew on deck to lift the boat evenly by hauling on the falls.*

"*Avast*": *cease hauling.*

"*Make fast*": *deck crew secures and coils down the falls and boat lines making the boat fast.*

Commands to oarsmen:

"Stand by oars": oarsmen place oarlocks, distribute oars from storage in the boat to each oarsman, who lays the oar along the gunwale of the boat, blade forward.

"Out oars": oarsmen lift oars in unison, swing them perpendicular to the gunwales and parallel to the water, and place them in their oarlocks; or, alternately—

"Toss oars": oarsmen lift oars to the vertical in unison, resting the inboard end on the floor of the boat between their legs.

"Let fall oars": oars are lowered smartly to the rowing position, as in "out oars."

"Give way together": oarsmen row in unison, each timing his stroke by keeping pace with the man in front of him.

"Hold water": securing themselves in the boat, oarsmen brace their oars in the water to retard the boat's progress.

"Back water" or "stern all": oarsmen row in unison in the opposite direction thereby moving the boat astern. At the command "back starboard (or larboard)" oarsmen on the side indicated back water.

"Oars": at the end of the present stroke, oarsmen bring their oars horizontal, as in "out oars."

"Way enough": oarsmen lift their oars out of the oarlock and place them along the gunwale, blade forward, returning them to the position assumed at "stand by oars." Oarlocks are secured.

"Boat oars": oarsmen pass their oars overhead, blade forward, and stack them in the center of the boat.

When operating in proximity to piers, landing places, or a ship's side, several other commands will be found useful:

"Cast off": release the lines holding the boat to a ship or pier.

"Shove off": use the boat's oars or boathooks to push the boat away from a dock, shore or vessel.

"Pick up the stroke": oarsmen not already giving way (or backing) begin to do so on the succeeding stroke. This command is employed after clearing a restricted space.

"In bows": forward most oarsmen cease rowing, stow their oars, and take position in the bows to handle lines or boathooks.

"Fend off": use the boat's oars or boathooks to keep the boat from striking a dock, obstruction or vessel.

Each of the ship's boats is fitted with a mast and sail. If the officer commanding determines that sailing the boat would be more expeditious than rowing, after ordering "boat oars," his relevant commands will be "step the mast," "set sail," and later "furl sail," and "unstep the mast."

BOARDING THE ENEMY

THE QUARTER BILL

Each man aboard Constitution *will know his place and duty before battle thanks to the Watch, Quarter, and Station Bill, which indicates his appointed role for any evolution of the ship, e.g. battle, tacking and wearing, anchoring, &c. Each man's stations are indicated using his hammock number rather than his name, which eliminates having to rewrite the quarter bill with every change in the crew. Although* Constitution's *crew numbers closer to 450, Mr. Blunt provides a proportional example of dispositions for battle in the Quarter Bill appearing below.*

A Quarter Bill.

For a frigate of 44 guns, with a compliment of three hundred men.

Quarter deck.

The captain to command the whole,	1
1st lieutenant to assist in fighting the ship,	1
Two midshipmen as aid de camps,	2
The [sailing] master to attend the steerage,	1
A master's mate to attend the braces and working ship,	1
A midshipman to fight the quarter deck guns,	1
Boatswain's mate and six men to repair damages,	7
Two quarter masters and two men at the helm,	4
A quarter master at the conn,	1
Seven men to each gun	35

A hand at the hatchway to receive powder,	1
Boys as carriers,	3
Purser and captain's clerk to take minutes	2
	60

Fore-castle.

A master's mate to command,	1
Boatswain and four men to repair damages,	5
Boatswain's mate [to] assist in working ship,	1
Seven men to each gun,	14
A boy as carrier,	1
A hand to give powder from the hatchway	1
	23

Main deck.

Second lieutenant and a midshipman foremost division,	2
Third lieutenant and a midshipman aftermost division,	2
Boatswain's mates working ship,	2
Captains of the mast at the mast,	2
Seven men to each gun,	105
Boys as [powder] carriers,	7
Two men at each hatchway to receive powder	9
	126

Tops.

A midshipman and four men in the fore and main tops with small arms and two in the mizzen,	12
Carpenter, his mate and two men in the wings	4

Magazine.

Gunner, his mate and two men,	4
Light room, the cook,	1
Master at arms, ship's corporal and six men in the 'tween decks to haul powder	8
	13

Cock-pit

Surgeon and his assistant,	2
Chaplain and two loblolly boys	3
	5

Marines.

First and second lieutenant,	2
A Sergeant and two corporals,	3
A drummer and fifer,	2
Rank and file	50
	57

Sum total 300

N.B. The marines should be stationed on the quarter deck, fore-castle and gangway, with musketry.

The calculation for regulating the men at quarters differs from that of all other stations, by including the marines, boys of the 1st class, and idlers. The boys of the 1st class may be entrusted to handle powder along the passages. The cooper should be quartered in the magazine, to be ready to open powder barrels. The ship's cook and master-at-arms in the light room. . . . The carpenters in frigates should assist, when not occupied in their own line, to hand powder between decks. The assistance of the purser, to superintend the distribution of powder at the hatchways, and, to arrange and command the people stationed to hand it along, appears to be a much more eligible mode of employing his services than to station him in the cockpit. . . .

CONSTITUTION'S ARMAMENT:

Although classified as a 44, Constitution *carries more guns than her rating, as is typical of fighting ships. She has a total of fifty-two iron naval guns of two types: thirty 24-pound long guns on her Gun Deck, and twenty 32-pound carronades and two 24-pound carronades (called gunnades) on her Spar Deck.*

The 24-pounder fires high-velocity, long-range projectiles of that weight, while the carronade fires heavier but low-velocity and shorter-range ammunition. While the carronade has about a third of the effective range of the long gun, it also has about a third of the weight and uses a third of the powder charge, and thus may be mounted higher in the ship without endangering her stability or structure.

Carronades, also called "smashers," are very effective against sails and rigging and against enemy crews in close fights, while long guns can inflict damage at much greater range. Constitution, *carrying both, thus can fight effectively either close in or at a distance, giving her captain more favorable options when conducting a battle. British frigates also carry a combination of carronades and long guns, their long guns, however, being 18-pounders. This gives the American 44s a significant advantage in broadside weight and at range, as when USS* United States *defeated HMS* Macedonian *in October 1812.*

Constitution's *guns employ several kinds of shot, depending upon utility:*

Type of shot	Description	Used against
Round (solid)	Solid cannonballs	Hull, masts, yards, and spars
Bar	Whole or half round shot connected by an iron bar.	Sails and rigging
Chain	Round shot or sections of iron connected by chain links	Sails and rigging
Star	An iron ring fitted with five, 3 to 4-foot lengths of chain.	Sails and rigging
Grape	Smaller round shot packed in a canvas bag.	Rigging and ship's company
Canister	Musket-sized balls packed in a metal container.	Ship's company

CLEARING FOR ACTION:

At quarters, Constitution's *crew is organized into five divisions, each led by one of her lieutenants. The First, Second, and Third divisions each man ten guns (five per side) on the Gun Deck. The Fourth Division mans the guns on the Spar Deck at the forecastle, while the Fifth serves the guns on the quarterdeck at the after end of the Spar Deck. Three-quarters of the crew are directly employed serving the guns in the divisions, while the remainder command, supply the guns, maneuver the ship, and attend to injuries to vessel and crew. As a midshipman, you well may be called upon to command one of* Constitution's *long guns on the Gun Deck.*

Exercise of the Great Guns.

When the drum beats to arms, every officer, and man, should repair to his respective station. The officers will muster their men, see everything requisite for coming into action in its proper place, as it relates to their respective departments, and report the same when done to the 1st Lieutenant.

The marines will fall in on the quarter-deck, and after having been mustered and inspected, they will be placed in the most advantageous posts for musquetry, as the judgment of the commanding officer may direct.

As illustration, detailed below is the gun crew of a carronade during Constitution's *battle against HMS* Java *on December, 29th 1812:*

Forecastle Division
Officers of Division (stationed at no. 1 chase gun):
John Alwyne, Lieutenant
James Delany, midshipman
Thomas Coursey, quarter gunner

Gun No. 4:
Stephen Webb, First Captain
Joseph Hacock, Second Captain and First Boarder
Joshua Perry, First Sponger
Enos Bateman, First Loader
Solomon Jenkins, Second Sponger and Fireman
Nicholas Vixtram, Second Loader
Robert Carter, Train tackle and Second Boarder
James Ellis, Powder Passer

A carronade in position.

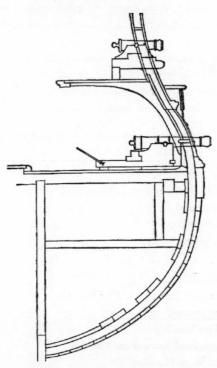

Section through a frigate-type ship showing the long guns on the main deck
and carronades on the spar deck.

A carronade's gun crew functions as follows:

Seven men to each gun with a boy to every two guns to carry powder, and each man ha[s] his allotted duty which prevent[s] all confusion. . . . The following is the duty . . . allotted to each man at the gun. Viz.

1st. The [gun] captain: He is to direct the men of his gun in all their evolutions. He is to point, prime, and fire the gun, being careful, the moment she is discharged, to stop the vent, with the twisted piece of oakum; a quantity of which, he should have for that purpose, in a small tin box with tubs for primers, buckled round his waist. He is also to prick the cartridge to know when it is home.

2nd. The rammer and sponger: He stands abaft the gun on the larboard side and before the gun on the starboard side, next to the muzzle. He is to sponge the gun and ram home the charge, and should be very careful in knocking off loose particles of fire from the sponge, on the sill of the port. When he has rammed home the charge, he should lay his sponge always athwart the deck, whilst he runs out the gun.

3rd. The powderman: He stands next to the muzzle, opposite to the sponger, he is to receive the charge and load the gun; he is to be careful in putting the cartridge in arse foremost and seam downwards; shot next to the cartridge and wad next to the shot; every third round he is to worm the gun.

4th and 5th are swabbers; they stand opposite each other next to the sponger and powder man; the one that stands aft, hands the cartridge and the wad to the powder man, and the one that stands forward hands the shot; they are also to swab round the gun, to take up loose particles of powder and assist the boy in fetching it.

6th and 7th are trainers; they stand opposite each other next to the breech, to train, elevate, or depress the gun as the captain directs.

Boy, fetches powder from the hatchway, and gives the cartridge-box to the swabber, and receives the empty one which he immediately hands below and receives another full one. The above regulations give three men to each tacklefall for running out the gun, and a captain to direct it, as they are killed, or wounded, the next takes his place, and when the gun is unmanned, the officers must remove men from other guns, make a virtue of necessity, and do the best they can.

Each gun crew is responsible for one gun on each side of the ship. The duplication of assignments, as with first and second captains, loaders, spongers, &c, permits a crew to fight both sides of the ship at once, as sometimes becomes necessary, and provides extra men to replace those felled in combat. In this instance, when Stephen Webb, first captain, was wounded severely during the fight against H.M.S. Java, he would have been replaced by the second captain, Joseph Hacock. Also several men within each gun crew are designated as:

Boarders, Sail-trimmers, and Firemen.

Boarders are the first division, under the command of the 1st Lieutenant, and are called on being boarded, to repel the attack;—this method not only prevents confusion, but gives every advantage, for whilst they are opposing the assailants, the remainder at the guns can keep up a brisk fire on the enemy, without having it returned, as he would be afraid of killing his own men; though should the boarders be found insufficient to defense the ship, the remaining divisions should be led in order, to the charge. . . . Boarding with the three divisions with pikes, cutlasses, and tomahawks, under their respective officers, and led by the captain, or commanding officer, with the marines to cover them, and hold the ground they gain . . . will speedily decide the contest, and spare a great effusion of blood.

Sail-Trimmers are the second division under the command of the 2d Lieutenant, and are to trim sails in action; they should also be subdivided into three divisions, as part of their duty may not require the whole, and the commanding officer can without confusion, order any part to the service required; for instance, "Send the first division of Sail-Trimmers aft to the weather main-brace. First division Sail-Trimmers, forward; second and third, aft, wear ship," &c. &c. when the duty is done they return to their quarters.

Boarding knife.

Firemen are the third division under the command of the 3d. Lieutenant:—they are always to have their fire-buckets kept in good order and hung in their places, as also near their respective quarters in action. They are to be subdivided into three divisions, as the ship in action may be on fire in several places at once. The first of the subdivisions are to work the engine, and the second and third divisions draw and carry water to the place in their fire-buckets. On coming into action in the night, they are to get the fire lanterns from the officer who distributes them, and hang in them against the side immediately over the port, and when [the] exercise or action is over, return them.

These divisions should be frequently exercised in their relative duties, for practice makes perfect, and perfection is mostly attended with success, and sham-fights both amuse and instruct crews. . . .

ENGAGING THE ENEMY:

Victory in war at sea is most certain to be gained by fast and accurate gunfire, made effectual by adroit maneuvering of the vessel. The nature of the contest, close in or at distance, will be dictated by which vessel holds the weather gauge, the weaponry that each possesses, and the abilities of captains and crews to gain advantage by maneuver. Since a ship is without the protection of framing in her bow and stern, and since shots delivered along the length of a ship are significantly more destructive than those fired athwart the vessel, the ideal maneuver is to pass ahead or astern of one's opponent, delivering a broadside at right angles. Besides the advantages aforesaid, "raking" also exposes the attacker to little fire from the enemy.

Failing such a decisive maneuver, a captain with an advantage in long guns would choose to remain at a distance from his opponent and attempt to disable his rigging and thereby his ability to maneuver. A captain with a preponderance of smashers, however, would probably attempt to close with the enemy ship to inflict the most damage on her hull and company. A third option, given an advantage or an equivalence in men, is to close-up to grapple and board the enemy. Mr. Blunt elucidates some of the particulars:

On Coming into Action.

When the Captain . . . deems it expedient, he will order the 1st Lieutenant to prepare for action.

The Boatswain, from the 1st Lieutenant's order, turns the hands up, to clear ship, which his mates repeat at each hatchway.

The sail-trimmers will trim sails and attend upon deck whilst in chase, get ... spare rigging, &c. on deck, and see all clear for shortening sail.

The boarders will get the main deck guns cleared of all encumbrances, and ready for action; ... have shot and wads brought upon deck ... boarding pikes, cutlasses, small arms, &c. at hand ... decks wet and strewed with sand, &c.

The firemen will get their fire-buckets near their respective quarters, a tub filled with water and a bucket at each hatchway ... [with] fire-tubs, half filled with water, with two swabs to each, and placed at their respective guns....

The gunner will receive the keys of the magazine from the 1st Lieutenant, open the same, and send the men that are stationed there into it; he will make them pull off their shoes, and see that they have no knives, nails, buckles, iron, steel, &c. about them. They will prepare to hand powder to the carriers....

A gunner's mate ... will inspect the guns and see that they are supplied with everything necessary for action....

The Action.

The necessary preparations being completed, and the gunners, &c. ready at their respective stations to obey the order, the commencement of the action is determined by the mutual distance and situation of the adverse ship. The cannon being levelled in parallel rows from the ship's side, the combat then begins by a vigorous cannonade, accompanied with the whole effort of the great guns and small arms, as the captain must endeavor to bring his ship within the point blank range of a musket of his adversary, which is the most convenient distance, so that all his artillery may do effectual execution.

The method of firing in platoons, or volleys of cannon at once, appears inconvenient in the sea service, and perhaps should never be attempted unless in the battering of a fortification. The sides and decks of the ship, although sufficiently strong for all the purposes of war, would be too much shaken by so violent an explosion and recoil. The general rule observed on this occasion through the ship, is to load, fire, and sponge the gun, with all possible expedition, yet without confusion or precipitation. The captain of each gun is particularly enjoined to fire only when the piece is properly directed to its object, that the shot may not be fruitlessly expended. The lieutenants, who command the different batteries, traverse the deck to see that the battle is prosecuted with vivacity, and to exhort the men to their duty. The midshipmen second these injunctions, and give the necessary assistance whenever it may be required, at the guns committed to their charge. . . .

The havoc produced by a continuation of this mutual assault, may be easily conjectured by the reader's imagination. Battering, penetrating and splintering the sides and decks; shattering and dismounting the cannon; mangling and destroying the rigging; cutting asunder or carrying away the masts; piercing and tearing the sails, so as to render them useless; and wounding and killing the ship's company.

The comparative vigour and resolution of the assailant to effect these previous consequences in each other, generally determines their success or defeat; we may say generally, because the fate of the combat may sometimes be determined by some unforeseen accident, equally fortunate for the one and fatal to the other; such, for instance, as one of the ships falling in the situation of raking the other; that is to say, cannonading her on the stern or head, so as that the balls shall range the whole length of the decks of her adversary. This is one of the most dangerous incidents that can happen in a sea engagement. It is frequently called *raking fore and aft*, and is similar to what engineers term *enfilading*. . . .

No one must quit his post upon pain of death, and should any one happen to refuse obeying his officer, he ought to be put to death on the spot. . . .

CAPTAIN'S ORDERS TO THE MEN AT THEIR GUNS:

The strictest silence is to be observed by the men at their guns, and the Officers are to prevent any unnecessary noise and all confusion. Particular orders will be given in sufficient time for the manner of shotting the guns; It is most likely that 2 round shot will be ordered for the two first round[s] on the Gun deck, one round shot & one load grape for the two first rounds on the upper-deck. The captains of the guns are ordered not to attempt to put any other charge into their guns, than that which is expressly ordered, as the fatal consequence of loading the gun improperly has been too often experienced with the loss of their lives. The greatest attention is ordered to be paid in pointing and elevating the guns, as it is most necessary & essential to fire a well-directed shot and to fire often. . . .

When [the] order for the gun locks is given, the first captains are to go for their locks and place them on the Starboard Guns,—the second captains in the same manner are to place their locks on the Larboard guns and they are to have their respective guns cast loose and ready for action on either side of the ship or on both as the order may be given for that purpose. . . .

There being a fire bucket appointed for each gun on each deck and its opposite gun in the ship, the bucket men are to bring their respective buckets opposite to their guns on coming to action, and they are to be in readiness to break off from their guns to supply water at any part of the ship it is called for. . . .

The officers commanding at their different quarters, are directed to see the Decks wetted and sand or ashes strewed upon them, whenever the ship is going into action also that water in the proper casks is supplied for the people in action. . . .

The boarders who may be called upon either to board the enemy [or] to repel the enemy who may attempt to board the Ship are to be attentive to the drum, & upon the call of the long roll they are immediately to assemble on both gangways according to their Stations, from which place their Officers will lead them on.

BOARDING:

The U.S. Naval Regulations of 1814 require that "the following number of men at least, (exclusive of marines) are to be exercised and trained up to the use of small-arms, under the particular care of a lieutenant or master at arms: 44 gun ship, 75 men." *Mr. Blunt offers the following remarks upon the enterprise itself:*

Of Boarding.

Boarding is the art of approaching the ship of an enemy so near, that you can easily, and in spite of him, throw on board the grapplings, which are fixed on the lower yard-arms, at the forecastle, gangways, &c. for the purpose of being thrown into the enemy's ship, as soon as alongside, in order to confine the vessels together, and give the people an opportunity of getting on board, to carry the adverse ship sword in hand. . . .

On preparing to board, the marines will be drawn up to cover the boarders. The boarders will be drawn up in two distinct close columns, with pikes and cutlasses, &c. under the command of the 2d. and 3d. Lieutenants.

Two platoons of 20 men each, under the command of mates or midshipmen, will be drawn up with musketry to act as lines of reserve, or support to the columns. Two detachments of 10 men (these men should be selected for good marksmen) each, with small arms under the command of midshipmen, will be drawn up to act as light companies or scouts, that is, to hover about and [pick] off the stragglers, men in the tops, &c.

When the word is given to board, the 1st. Lieutenant is to command in person.

The marines will open a brisk fire on the enemy's deck to afford a landing. The column commanded by the 2nd. Lieutenant will board first and get possession of the opposite side of the deck.

The column under the command of the 3d. Lieutenant will follow and take possession of the side of the deck on which they board.

Both columns will then charge the enemy along each gangway, and when they gain the quarter-deck each column will extend its line by bringing up the rear ranks in front; however, they are not to be less than three deep.

The lines of reserve follow next; they are to support and flank the columns, the detachments then go over, they are to hover about, pick the men off, in the tops, waist, and wherever they can see them, aiming chiefly at the officers; in short they are the same as light or rifle companies in a field of battle. The marines will remain in the rear some distance, and keep possession of the ground already gained, and those in the waist in check.

The 1st. Lieutenant's judgement and circumstances must now certainly direct him, and should he get repulsed, he should endeavor to rally his men, which if found impossible he should conduct a good retreat, under the cover of the lines of reserve and marines.

When the enemy has struck, her commanding, or surviving officer, should be brought on board to surrender his sword, the ship's papers secured, the prize manned, and the prisoners removed, and well guarded, the guns of both ships, and magazines should be next secured, the dead thrown overboard, and the decks washed down, and everything safely returned to its respective departments; after which, a general muster should take place, and a list of the killed, wounded and missing taken, after which, both ships will repair damages, and either separate or remain in company, as the commander may judge expedient.

Constitution *is authorized to carry a force of sixty marines: two officers, six petty officers, a fifer and drummer, and fifty privates. In battle, marines occupy positions on deck and in the tops. With their musquetry they attack exposed officers, support boarding parties, and repel boarders. They may also serve guns or in landing parties. The small arms used in boarding or defense include muskets, pistols, swords, pikes, and axes. Mr. Blunt offers insightful remarks upon two of the weapons most effective for boarding.*

The Pike.

The pike is the most useful weapon used on board ships of war, either in attack or defense, as a close column of pikes formed in order, with musquetry in their rear, present such a barrier to the enemy that they can seldom or ever cut through, or sustain the force and weight of this phalanx in a well-directed vigorous charge. . . .

The Sword.

Few men require a more perfect knowledge of the sword than seamen of armed vessels, and unfortunately few possess it less. In action a good swordsman has not only a better chance of preserving his own life, but is more destructive to an enemy . . . for with judicious parries and scientific cuts and thrusts he mows down all the inexperienced that oppose him. . . .

Thus are battles won at sea, by the iron fortitude and zealous devotion to duty of those who fight the ship.

OF PRIZES AND PENSIONS

One incentive for men such as yourself to join the colors and to fight bravely and well is the potential for prize money earned through the auction of captured ships and their stores. This one of your fellow midshipmen declares "the golden fruit, which is much the sweetest part of Warfare." *Senior midshipmen are often assigned to command prizes upon capture and to sail them to friendly or neutral ports for disposition.*

RULES GOVERNING PRIZES
In 1800, in An Act for the Better Government of the Navy of the United States, *the U.S. Congress stipulated the procedures for establishing a prize claim and against the removal of property or mistreatment of prisoners therefrom. The same law also established the rules determining the value of an award, and:*

That the prize money, belonging to the officers and men, shall be distributed in the following manner:

I. To the commanding officers of fleets, squadrons, or single ships, three twentieths, of which the commanding officer of the fleet or squadron shall have one twentieth, if the prize be taken by a ship or vessel acting under his command, and the commander of the single ships, two twentieths; but where the prize is taken by a ship acting independently of such superior officer, the three twentieths shall belong to her commander.

II. To sea lieutenants, captains of marines, and sailing masters, two twentieths; but where there is a captain, without a lieutenant of marines, these officers shall be entitled to two twentieths and one third of a twentieth, which third, in such case, shall be deducted from the share of the officers mentioned in article No. III. of this section.

III. To chaplains, lieutenants of marines, surgeons, pursers, boatswains, gunners, carpenters, and master's mates, two twentieths.

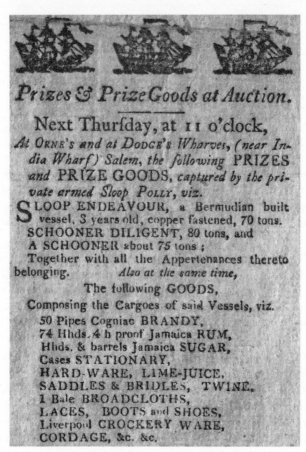

© USS Constitution Museum Collection.

IV. To midshipmen, surgeon's mates, captain's clerks, schoolmasters, boatswain's mates, gunner's mates, carpenter's mates, ship's stewards, sail-makers, masters at arms, armorers, coxswains, [cooks,] and coopers, three twentieths and a half.

V. To gunner's yeomen, boatswain's yeomen, quartermasters, quarter-gunners, sail-maker's mates, sergeants and corporals of marines, drummers, fifers and extra petty officers, two twentieths and [a] half.

VI. To seamen, ordinary seamen, marines, and all other persons doing duty on board, seven twentieths. . . .

Awards to *Constitution's* Officers and Crew for the Ship's Victories of 1812:
Since both HMS Guerriere *and HMS* Java *were destroyed at sea after their captures, no prizes exist upon which to make a claim. However, the government understands the importance of prize money, both as fair reward for risks assumed and as an inducement to naval service. In his letter of November 21, 1812 to the chairman of the naval committee of the U.S. House of Representatives, Secretary of the Navy Paul Hamilton wrote:*

Had Capt. Hull . . . succeeded in getting the *Guerriere* into Port, the officers and crew [of] *Constitution*, considering the *Guerriere* as her equal, would have been entitled to the whole of the *Guerriere*, her stores and prize goods, sooner however than run the risk of losing the *Constitution*, he determined to destroy the whole. The question then arises, what, under these circumstances, ought the officers and crew to be allowed? For my own part, I have no hesitation in offering it as my opinion, that the sum of one hundred thousand dollars, would not be too liberal a provision, or too great an encouragement for the great gallantry, skill, and sacrifice of interest displayed on the occasion and am persuaded that if such a provision were made, the difficulties of manning our frigates, at present experienced, would vanish.

In this judgment the President of the United States concurred, writing:
To the Senate and House of Representatives of the United States.

I lay before Congress a letter, with accompanying documents, from Captain Bainbridge, now commanding the United States frigate "the Constitution," reporting his capture and destruction of the British frigate "the Java." The circumstances and the issue of this combat, afford another example of the professional skill and heroic spirit which prevail in our naval service. The signal display of both, by Captain Bainbridge, his officers, and crew, command the highest praise.

This being a second instance in which the condition of the captured ship, by rendering it impossible to get her into port, has barred a contemplated reward of successful valor, I recommend to the consideration of Congress, the equity and

———— ☸ ————

propriety of a general provision, allowing, in such cases both past and future, a fair proportion of the value which would accrue to the captors, on the safe arrival and sale of the prize.

JAMES MADISON

February 22d, 1813.

To this communication the Congress responded with:

AN ACT Rewarding the Officers and Crew of the Frigate *Constitution*. . . .

Be it enacted, &c., That the President of the United States be, and he is hereby, authorized to have distributed, as prize money, to Captain Isaac Hull, of the frigate *Constitution*, his officers and crew, the sum of fifty thousand dollars, for the capture and destruction of the British frigate *Guerriere*: and the like sum, in like manner, to Captain William Bainbridge, his officers and crew, for the capture and destruction of the British frigate *Java*. . . .

Approved, March 3, 1813.

Under this Act, the $50,000 awarded by Congress for the victory over HMS Guerriere (or the like sum for the victory over HMS Java), means that a share constituting one-twentieth of the total prize equals $2,500. Under the 1800 Act the total prize is divided thus:

Category	No. Shares	Total Value	No. Persons	Indiv. Award	Years' Pay
I. Commander	3	$ 7,500	1	$7,500	6.25
II. Lieuts, etc.	2	$ 5,000	8	$ 625	1.3
III. Chaplain, &c.	2	$ 5,000	10	$ 500	0.83
IV. Mdshpmn &c.	3.5	$ 8,750	30	$ 291.67	0.61
V. Yeomen, &c.	2.5	$ 6,250	25	$ 250	0.91
VI. Seamen, &c.	7	$17,500	364	$ 48.08	0.33
Total	20	$50,000	438		

Pensions Awarded for Naval Service:

An Act for the Better Government of the Navy of the United States *(1800)* *also established a pension system for naval service, with awards to be paid from the government's share of prize monies as follows:*

That all money accruing, or which has already accrued to the United States from the sale of prizes, shall be and remains forever a fund for the payment of pensions and half pay, should the same be hereafter granted, to the officers and seamen who may be entitled to receive the same; and if the said fund shall be insufficient for the purpose, the public faith is hereby pledged to make up the deficiency; but if it should be more than sufficient, the surplus shall be applied to the making of further provision for the comfort of the disabled officers, seamen, and marines, and for such as, though not disabled, may merit by their bravery, or long and faithful services, the gratitude of their country.

Awards to Relations of Those Killed in the Line of Duty:

An Act for Providing for Navy Pensions

In certain cases.

If any officer of the navy or marines shall be killed or die, by reason of a wound received in the line of his duty, leaving a widow, or if no widow, a child or children, under sixteen years of age, such widow, or if no widow, such child or children, shall be entitled to receive half the monthly pay to which the deceased was entitled at the time of his death, which allowance shall continue for and during the term of five years: but in case of the death or intermarriage of such widow, before the expiration of the said term of five years, the half pay for the remainder shall go to the child or children of the said deceased officer: provided, that such half pay shall cease on the death of such child or children: and the money required for this purpose shall be paid out of the navy pension fund, under the direction of the commissioners of that fund.

Awards to those Disabled in the Line of Duty:

From: An Act Providing a Naval Armament, approved 1 July 1797:

Sec. 11. *And be it further enacted,* That if any officer, non-commissioned officer, marine or seaman belonging to the navy of the United States, shall be wounded or disabled while in the line of his duty in public service, he shall be placed on the list of the invalids of the United States, at such rate of pay and under such regulations as shall be directed by the President of the United States: *Provided always*, That the rate of compensation to be allowed for such wounds or disabilities to a commissioned or warrant officer shall never exceed for the highest disability half the monthly pay of such officer at the time of his being so disabled or wounded; and that the rate of compensation to non-commissioned officers, marines and seamen, shall never exceed five dollars per month: And provided also, That all inferior disabilities shall entitle the person so disabled to receive an allowance proportionate to the highest disability.

"OLD IRONSIDES": HER EXPLOITS OF 1812

Watching shot bounce harmlessly off her sides during her fight against HMS Guerriere, one of Constitution's sailors yelled, "Huzzah, her sides are made of iron!" Thus was established a new name for a vessel both fortunate and glorious. Study her achievements and find pride and inspiration from her successes in that signal year of 1812. As one of her midshipmen, you must do your best, if the chance offers, to help her add to her laurels.

THE DECLARATION:

AN ACT,

Declaring War between the United Kingdom
of Great Britain and Ireland and the dependencies thereof,
and the United States of America and their Territories.

BE it enacted by the Senate and House of Representatives of the United States of America in Congress assembled, That war be and the same is hereby declared to exist between the United Kingdom of Great Britain and Ireland, and the dependencies thereof, and the United States of America, and their territories; and that the President of the United States be, and he is hereby authorized to use the whole land and naval force of the United States to carry the same into effect, and to issue to private armed vessels of the United States commissions or letters of marque and general reprisal, in such form as he shall think proper, and under the seal of the United States, against the vessels, goods, and effects of the government of the said United Kingdom of Great Britain and Ireland, and the subjects thereof.

H. Clay,
Speaker of the House of Representatives.
William H. Crawford,
President of the Senate *pro tempore*.
June 18, 1812.—Approved,
James Madison

THE CHASE:

U.S. Frigate *Constitution*

We have the pleasure of announcing the arrival in our harbor, the 26th [of July, 1812] of the frigate *Constitution*, Capt. [Isaac] Hull. She left the Chesapeake Bay on the 12th inst. and on the 16th in the afternoon, saw a frigate, and gave chase; the winds being light they could not come near enough to ascertain who she was. It continued calm the principal part of the night. On the morning of the 17th, an English squadron was discovered, consisting of a ship of the line, four frigates, a brig and schooner;—the nearest frigate within gun shot. Throughout the whole of this day it was calm; and every exertion made, by towing and [kedging] to make head way; but the enemy by attaching all their boats to two of the frigates, were evidently gaining upon the *Constitution*, and occasionally enabled them to

SALEM,

TUESDAY, SEPTEMBER 1, 1812.

Brilliant Naval Victory!

The United States frigate Conſtitution, Captain HULL, anchored on Sunday in Boſton harbor, from a ſhort cruiſe, during which ſhe fell in with the Engliſh frigate Guerriere, which ſhe captured, after a ſhort, but ſevere action — The damage, ſuſtained by the fire of the Conſtitution, was ſo great, that it was found impoſſible to tow her into port, and accordingly the crew were taken out, and the ſhip burnt. The brilliancy of this action will excite the livelieſt emotions in every American boſom.

© USS Constitution Museum Collection.

bring their bow guns to bear upon her. This kind of maneuvering, and the frequent discharge of the *Constitution*'s stern chasers, continued during the whole of this day, on the 18th at day light, a small breeze sprang up, when the *Constitution* spread all her canvas, and by out-sailing the enemy, escaped a conflict, which she could not have maintained with any hope of success against a force so greatly superior. The chase was continued 60 hours, during which time the whole crew remained at their stations. The *Constitution* was bound to New York, but from the unfavorableness of the winds, has put in here.

—*The Weekly Messenger* (Boston), 31 July 1812.

THE DEFEAT OF HMS *GUERRIERE:*

Captain Isaac Hull to Navy Secretary Paul Hamilton:

US Frigate *Constitution*
Off Boston Light, August 28th 1812

Sir,

I have the honour to inform you that on the 19th inst. at 2 pm, being in Latitude 41°. 42' Longitude 55°. 48' with the wind from the Northward, and the *Constitution* under my command steering to the S.SW. a sail was discovered from the mast head bearing E. by S. or E.SE. but at such a distance that we could not make out what she was. All sail was immediately made in chase, and we soon found we came fast up with the chase, so that at 3 PM we could make her out to be a ship on the starboard tack close by the wind under easy sail. At ½ past 3 PM, closing very fast with the chase [we] could see that she was a large frigate, At ¾ past 3 the chase backed her maintopsail, and lay by on the starboard tack; I immediately ordered the light sails taken in, and the royal yards sent down, took two reefs in the topsails, hauled up the foresail, and mainsail and [saw] all clear for action; after all was clear the ship was ordered to be kept away for the enemy, on hearing of which the Gallant crew gave three cheers, and requested to be laid close alongside the chase. As we bore up she hoisted an English ensign at the mizzen gaff, another in the mizzen shrouds, and a jack at the fore, and mizzen topgallant mast heads. At 5 minutes past 5 PM, as we were running down on her weather quarter, she fired a broadside, but without effect, the shot all falling short; she then wore and gave us a broadside from

[her] larboard guns, two of which shot stuck us but without doing any injury. At this time, finding we were within gunshot, I ordered the ensign hoisted at the mizzen peak, and a jack at the fore and mizzen topgallant mast head, and a jack bent ready for hoisting at the main; the enemy continued wearing and maneuvering for about ¾ of an hour, to get the wind of us. At length finding that she could not, she bore up to bring the wind, on the quarter, and run under her topsails, and jib; finding that we came up very slow, and were receiving her shot without being able to return them with effect, I ordered the main topgallant sail set, to run up alongside of her.

At 5 minutes past 6 PM being alongside, and within less than Pistol Shot, we commenced a very heavy fire from all of our Guns, loaded with round and grape, which [did] great execution, so much so that in less than fifteen minutes from the time we got alongside, his mizzen mast went by the board, and his main yard in the slings, and the hull, and sails very much injured, which made it very difficult for them to manage her. At this time the *Constitution* had received but little damage, and having more sail set than the enemy she shot ahead; on seeing this I determined to put the helm to port, and oblige him to do the same or suffer himself to be raked; but our getting across his bows, on our helm being put to port the ship came to, and gave us an opportunity of pouring in upon his larboard bow several broadsides, which made great havoc amongst his men on the forecastle and did great injury to his fore rigging, and sails. The Enemy put his helm to Port, at the time we did, but his mizzen mast being over the quarter, prevented her coming to, which brought us across his bows, with his bowsprit over our stern. At this moment I determined to board him, but the instant the boarders were called, for that purpose, his foremast, and mainmast went by the board, and took with them the jib-boom, and every other spar except the bowsprit. On seeing the enemy totally disable[d], and the *Constitution* received but little injury, I ordered the sails filled, to haul off and repair our damages and return again to renew the action, not knowing [whether] the Enemy had struck or not, we stood off for about half an hour to repair our braces and such other rigging as had been shot away, and wore around to return to the Enemy; it being now dark, we could not see whether she had any colours flying or not, but could discover that she had raised a small flag staff or jurymast forward. I ordered a boat hoisted out, and sent Lieutenant Reed on board as a flag to see whether she had surrendered or not, and if she had, to see what assistance she wanted, as I believed she was sinking. Lieutenant Reed returned in about twenty minutes, and brought with him James Richard Dacres, Esqr, commander of his Britannic Majesty's frigate the *Guerriere*, which ship had surrendered to the United States frigate *Constitution*; our boats were immediately hoisted out and sent for the prisoners, and were kept at work bringing them and their baggage on board, all night. At daylight we found the enemy's ship a perfect wreck, having many shot holes between wind and water, and above six feet of the plank below the bends taken out by our round shot, and her upperwork[s] shattered to pieces, that I determined to take out the sick and wounded as fast as possible, and set her on fire, as it would be impossible to get her into port.

At 3 PM all the prisoners being out, Mr. Reed was ordered to set fire to her in the store rooms, which he did and in a very short time she blew up. I want words to convey to you the bravery and gallant conduct, of the officers and the crew under my command during the action. I can therefore only assure you, that so well directed was the fire of the *Constitution*, and so closely kept up, that in less than thirty minutes, from the time we got alongside of the enemy (one of their finest frigates) she was left without a spar standing, and the hull cut to pieces, in such a manner as to make it difficult to keep her above water, and the *Constitution* in a state to be brought into action in two hours. Actions like these speak for themselves which makes it unnecessary for me to say anything to establish the bravery and gallant conduct of those that were engaged in it. Yet I cannot but make you acquainted with the very great assistance I received from that valuable officer Lieutenant Morris in bringing the ship into action, and in working her whilst alongside the enemy, and I am extremely sorry to state that he is badly wounded, being shot through the body. We have yet hopes of his recovery, when I am sure, he will receive the thanks, and gratitude of his county, for this and the many gallant acts he has done in its service.

Were I to name any particular officer as having been more useful than the rest, I should do them great injustice, they all fought bravely, and gave me every possible assistance that I could wish. I am extremely sorry to state to you the loss of Lieutenant Bush of Marines. He fell at the head of his men in getting ready to board the enemy. In him our country has lost a valuable and brave officer. After the fall of Mr. Bush, Mr. Contee took command of the Marines, and I have pleasure in saying that his conduct was that of a brave good officer, and the Marines behaved with great coolness, and courage during the action, and annoyed the enemy very much whilst she was under our stern. . . .

I have the honour to be, with very great respect, Sir, Your Obedient Servant,

Isaac Hull

Captain James Dacres, R.N., to Vice-Admiral Herbert Sawyer, R.N.:
Boston, 7ᵗʰ September 1812

Sir,

I am sorry to inform you of the capture of His Majesty's late ship *Guerriere* by the American frigate *Constitution* after a severe action on the 19th of August in Latitude 40.20 N and Longitude 55.00 West. . . .

The loss of the ship is to be ascribed to the early fall of the mizzen mast, which enabled our opponent to choose his position; I am sorry to say we suffered severely in killed and wounded and mostly whilst she lay on our bow from her grape and musketry; in all 14 killed and 63 wounded, many of them severely. . . . The Enemy had such an advantage from his Marines and Riflemen when close, and his superior sailing enabled him to choose his distance. . . .

Signed J R Dacres

THE DEFEAT OF HMS *JAVA*

Extract of journal of Commodore Bainbridge
on board Frigate *Constitution*.

Tuesday 29th December 1812

At 9 AM, discovered two strange sails on the weather bow, at 10 AM discovered the strange sails to be ships, one of them stood in for the land, and the other steered off shore in a direction towards us. At 10.45 we tacked ship to the Nd & Wd and stood for the sail standing towards us. At 11 tacked to the Sd & Ed hauled up the mainsail and took in the royals. At 11.30 AM made the private signal for the day which was not answered, & then set the mainsail and royals to draw the strange sail off from the neutral coast.

. . . . Clear weather and moderate breezes from E.N.E. Hoisted our ensign and pendant. At 15 minutes past meridian, the ship hoisted her colours, an English ensign. . . . At 1.26 being sufficiently from the land, and finding the ship to be an English frigate, took in the main sail and royals, tacked ship and stood for the enemy.

At 1.50 P.M. the enemy bore down with an intention of raking us, which we avoided by wearing. At 2, P.M. the enemy being within half a mile of us and to windward, & having hauled down his colours . . . induced me to give orders to the officer of the 3rd Division to fire one gun ahead of the enemy to make him show his colours, which being done brought on afire from us of [our] whole broadside, on which he hoisted an English ensign . . . and then immediately returned our fire, which brought on a general action with the round and grape.

The enemy kept at a much greater distance than I wished, but could not bring him to closer action without exposing ourselves to several rakes. Considerable manoeuvers were made by both vessels to rake and avoid being raked.

. . . . At 2.10 P.M, commenced the action within good grape and canister distance. The enemy to windward (but much farther than I wished). At 2.30. P.M. our wheel was shot entirely away. At 2.40. determined to close with the enemy, notwithstanding her raking, set the fore sail & luffed up close to him. At 2.50, the enemy's jib boom got foul of our mizzen rigging. At 3 the head of the enemy's bowsprit & jib boom shot away by us. At 3.[0]5 shot away the enemy's foremast by the board. At 3.15 shot away the enemy's main top mast just above the cap. At 3.40 shot away gaff and spanker boom. At 3.55 shot his mizzen mast nearly by the board. At 4.[0]5, having silenced the fire of the enemy completely and his colours in main rigging being [down], supposed he had struck. Then hauled about the courses to shoot ahead to repair our rigging, which was extremely cut, leaving the enemy a complete wreck; soon after discovered that the enemy's flag was still flying; hove to to repair some of our damages. At 4.20, the enemy's main mast went by the board. At 4.50 [wore] ship and stood for the enemy. At 5.25 got very close to the enemy in a very [effective] raking position athwart his bows & was at the very instance of raking him, when he most prudently struck his flag.

Had the enemy suffered the broadside to have raked him previously to striking, his additional loss must have been extremely great; laying like a log upon the water, perfectly unmanageable, I could have continued raking him without being exposed to more than two of his guns (if even them).

After the enemy had struck, wore ship and reefed the top sails, hoisted out one of the only two remaining boats we had left out of 8, & sent Lieut. Parker, 1st [lieutenant] of the Constitution, on board to take possession of her, which was done about 6 P.M. The action continued from the commencement to the end of the fire, 1 h., 55'; our sails and rigging were shot very much, and some of our spars injured—had 9 men killed and 26 wounded. At 7 P.M. the boat returned from the prize with Lieut. Chads, the 1st of the enemy's frigate (which I then learnt was the *Java* rated 38—had 49 Guns mounted) .... Capt. Lambert of the Java was too dangerously wounded to be removed immediately.

.... The *Java* ... must have had upwards of 400 souls, she had one more man stationed at each of her guns on both decks than what we had. The enemy had 83 wounded & 57 killed.... Everything was blown up, except the officers' baggage, when we set her on fire on the 1st of January 1813 at 3 P.M. Nautical Time.

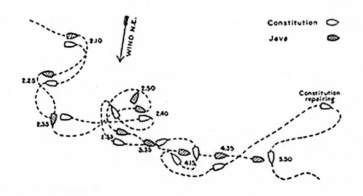

Excerpts from a letter from Henry D. Chads, First Lieutenant and senior survivor of HMS Java, *to Secretary of the Admiralty John W. Croker, written aboard* Constitution, *Dec. 31st, 1812:*

Sir,

.... On the morning of the 29th inst. at 8 AM off St Salvador (Coast of Brazil) the wind at NE we perceived a strange sail, made all sail in chase and soon made her out to be a large frigate ... at 2:10 when about half a mile distant she opened her fire giving us her larboard broad-side which was not returned till we were close on her weather bow; both ships now maneuvered to obtain advantageous positions; our opponent evidently avoiding close action and firing high to disable our masts in which he succeeded too well having shot away the head of our bowsprit with the jibboom and our running rigging so much cut as to prevent our preserving the weather gage. . . .

I cannot conclude this letter without expressing my grateful acknowledgement thus publicly for the generous treatment Captain Lambert and his officers have experienced from our gallant enemy Commodore Bainbridge and his Officers. . . .

HEROES OF USS *CONSTITUTION*

Of the reception for Captain Isaac Hull upon returning to Boston:

When Captain Hull landed from the *Constitution*, he was received with every demonstration of affection and respect. The Washington Artillery posted on the wharf again welcomed him with a federal salute, which was returned from the *Constitution*. An immense assemblage of citizens made the welkin ring with loud and unanimous huzzahs, which were repeated on his arrival in State Street, and at the coffee house. The street was beautifully decorated with American flags.—*Niles Weekly Register*, Sept. 12th, 1812.

Of Lieutenant William S. Bush of Marines, felled in the fight against HMS Guerriere:

Boston
September 13th, 1812

Wm. Lewis Bush
  Sir

I received yours of the 8th & hope you will pardon my negligence in not dating my letter & more particularly in not describing the death of your Gallant Brother which I thought I had done but which I now will do.

In the heat of the action the Marines were called aft; led on by the illustrious Bush, who, mounting the taffrail, sword in hand, and as he exclaimed "Shall I board her" received the fatal ball. . . . Thus fell that Great and Good officer who, when living was beloved, & now gone is lamented by all.

His loss is deeply regretted by his country & friends, but he died as he lived, with honor to both. . . .

With sentiments of the highest respect

I remain Sir

> your most obt & most hble
> Servant
> John Contee
> Lt. Marines

Of Lieutenant John Cushing Aylwin, wounded mortally in the fight against HMS Java:

From the Journal of Amos A. Evans, Surgeon, U.S.N.

January 29, Friday.—About one o'clock this morning Lieut. John Cushing Aylwin died of a malignant intermittent [fever] caused by a wound thro' the shoulder received in the action with the *Java*. A braver or better man never lived. His country has suffered an irreparable loss in the death of this young man. His many virtues have endeared him to the hearts of all who had the pleasure of his acquaintance, particularly his messmates. He bore his pain with great fortitude & was resigned to his fate, observing that he had witnessed death in too many shapes to be alarmed at his approach. . . . In the evening his body was committed to the deep with the honors of war. Lat. Ob: 16° 29' N. Long. 52° 30' W.

As a midshipman, you should resolve to emulate the valor, character, and utility of these men.

British Reaction to the American Victories

The public will learn with sentiments, which we shall not presume to anticipate, that a *third* British frigate has struck to an American. . . . This is an occurrence that calls for serious reflection—this, and the fact stated in our paper, of yesterday, that Lloyd's lists contains notices of upwards of five hundred British vessels captured, in seven months, by the Americans. *Five Hundred merchantmen*

and three Frigates. . . . Anyone who had predicted such a result of an American war, this time last year, would have been treated as a madman or a traitor. . . . Yet down to this moment, not a single American frigate has struck her flag . . . nothing chases, nothing intercepts, nothing engages them but to yield to them in triumph. . . .

—The Times, *London, Saturday, March 20, 1813*

A NAVAL AND NAUTICAL GLOSSARY

aback: sails forced aft against a mast by the wind, as when tacking.

abaft: aft of.

able seaman: a sailor rated more proficient than an ordinary seaman.

aft (after): toward the stern of the ship.

afterguard: those assigned to duty on the quarterdeck during maneuvers.

aloft: in reference to the masts and rigging of a ship, or into the air as with a rocket.

amidships: toward the middle of the ship from either fore or aft or from the sides.

anchor: a heavy device used to secure a ship or boat to the seabed;

> *bower:* carried at the ship's bow and normally used to anchor;

> *kedge:* relatively light, used for kedging;

> *sheet:* large, carried in the waist of the ship, and used to hold a ship in emergencies, such as to keep from being driven onto a lee shore in a storm.

avast: a command to cease any activity.

ballast: weight carried low in a ship to improve her stability.

beam: the maximum width of a ship; the side of the ship between the bow and quarter (abeam).

beat to quarters: to be called to battle stations by the Marine drummer.

between wind and water: the portion of the hull below the waterline temporarily exposed by a ship's movement in the sea, and thus vulnerable to enemy fire.

billet head: bow ornamentation other than a figurehead, usually a scroll.

binnacle: the supporting structure adjacent to the helm that houses the ship's compass.

boatswain: the warrant officer in charge of rigging; stationed in forecastle for both battle and repair; also chief of the crew, responsible for summoning it for various evolutions.

boom: a horizontal spar affixed at its forward end to a mast, to which the foot of a fore-and-aft sail is attached; re-rigged a boom can serve as a derrick to work cargo, etc.

bow: the forward end of a ship. Also starboard and larboard bows, indicating the areas to 45 degrees of either side of dead ahead.

bowsprit: a large spar resting on the stem and extending forward from the ship's hull to which the forestays and foresails are fastened.

breasthook: a thick, curved timber fastened across the inside of the stem to hold the bow and sides together.

broadside: the nearly simultaneous firing of all of the guns on one side of a ship; the total weight of metal launched by such action.

butts: casks used to hold wine, beer or water.

camboose (or caboose): the galley; the cook's stove.

capstan: a revolving cylinder on a vertical axis used to heave in lines or raise heavy weights such as anchors or cargo, the work being supplied by sailors pushing stout wooden bars inserted horizontally into the capstan head.

carronade: a low-velocity, short-range cannon first produced by the Carron iron works at Falkirk, Scotland. Its advantages are that it is lighter and requires a smaller crew, thus more may be used in a given space, increasing broadside weight. (Also called "smashers.")

cashier: to expel an officer from the service dishonorably.

ceiling: the inner planking covering the ship's frames.

chains: a platform extending from the side of the ship used to secure the shrouds and used by sailors when taking soundings.

chase guns (chasers): lighter guns carried at the bow or stern and fired when chasing or being chased.

cockpit: the area at the after end of the orlop deck used as an operating theater during battle.

company: the ship's entire complement of officers and men.

compass point: the division of the 360-degree circle of a compass card into 32 points, making each point equivalent to 11.25 degrees.

cutter: a ship's boat designed more for speed than capacity; features a square stern.

davits: small cranes used in pairs that are suspended outboard to lower or raise a ship's boats.

diagonal riders: interior braces extending from the keelson to the gun deck designed to stiffen the hull and to permit a greater weight of armament to be carried.

displacement ton: the weight of a vessel as measured by the weight of the water she displaces (in long tons of 2,240 pounds).

draft: the maximum depth of a ship below the waterline.

engine: the ship's fire pump.

ensign: a national flag flown by a warship from the gaff while underway, especially as identification before battle. Ensigns of other nations may be hoisted temporarily to deceive an enemy. When ships of the British squadron chasing *Constitution* off New Jersey in 1812 hoisted the U.S. ensign to try to lure into capture an American merchant ship, Captain Hull hoisted British colors to drive her away!

evolution: a tactical maneuver by a ship (e.g., tacking) or an exercise by the crew (e.g., gun drill).

falls: the system of tackle used to lower or raise a ship's boats.

firkin: a small cask used to hold liquids, butter or fish.

flemish: to coil a line down on a deck in a flat, tight, circle; done for both safety and appearance.

footrope: a rope attached to the underside of a yard upon which sailors stand to work sails. (Also called "horse.")

fore: toward the bow or forward part of the ship.

forecastle: a short raised deck at the forward end of a ship, or, in *Constitution*'s case, the forward end of her spar deck. (Pronounced "folk-sul.")

foremast: the forward-most of the ship's vertical masts in a vessel with two or more masts.

frames: the vertical wooden ribs that extend from the keel perpendicularly to the main deck to produce the shape of the hull, support the deck beams, and to which the outer (strakes) and inner (ceiling) hull planking is attached.

frigate: a full-rigged warship, smaller than a ship of the line, with one or two gun decks; used for scouting, communicating, or raiding in large fleets; the heavy raiding ship of the U.S. Navy in its early years.

full-rigged (or fully rigged): a ship with at least three masts, all square-rigged, though generally with fore-and-aft sails on her lower mizzen mast.

furl: to gather together and secure a sail.

gaff: the upper spar on a four-sided, fore-and-aft sail, such as the spanker hoisted on the mizzen mast. The ship's ensign normally flies from the spanker gaff.

gig: the ship's boat reserved for the use of the captain.

grog: rum or bourbon cut with an equal amount of water and issued to the crew as a daily spirit ration.

gunnade: a later development of the carronade mounted with a lower center of gravity.

gunwale: the upper edge of the side of a boat. (Pronounced "gun-ul.")

head: the bow.

heave: to throw; to haul (heave in); to stop a ship (heave-to).

hog: an upward vertical distortion of the center of a ship's keel due to the weight of the less buoyant bow and stern sections.

hove: see heave.

hove-to: a ship with her sails trimmed so that she is making no headway, generally with her bow into the wind; the maneuver is used to lower boats, effect repairs, or ride out rough weather.

idler: a crew member who stands no regular watches, but is subject to call at any hour, such as the surgeon or sailmaker.

in ordinary: an inactive ship kept in reserve but ready for use with relatively short notice.

inboard: toward the center of a ship.

jibboom: a spar extending the bowsprit to carry additional jib sails.

jib: a triangular fore-and-aft sail set forward of the foremast; a flying jib being set forward of a regular jib.

jolly boat: smallest of the ship's boats, used to transport persons or small items to or from the ship.

kedge: to move a ship by carrying one of her kedge anchors ahead with a boat, dropping it, and hauling up to it using the capstan, repeating the process alternately with her other kedge anchor. Used to move a ship during unfavorable winds or tides.

keel: the heavy, fore-and-aft wooden spine at the bottom of a ship to which the frames, stem, and stern are affixed.

keelson: timbers bolted atop the keel to strengthen it and, in *Constitution*'s case, provide a footing for her diagonal riders.

kentledge: flat, pig-iron ballast carried on the orlop deck to increase a ship's stability.

knot: A measure of speed equal to one nautical mile (approximately 6,080 feet) per hour. The term derives from the practice of measuring speed by throwing a log over the stern attached to the ship by a line with knots tied in it at exact intervals. The number of knots that pay out during a given time indicate the speed.

larboard: the left side of a ship when facing forward; to the left.

lee: in the opposite direction from which the wind comes (opposite: windward).

leeward: the direction toward which the wind blows; the downwind side of a ship (opposite: windward or weather side). (Pronounced "loo-urd.")

light room: a compartment adjacent to and separate from a magazine which displays behind glass the open-flame lantern used to illuminate the magazine.

line: the proper term for a rope when used aboard ship. With only rare exceptions (e.g., footrope), rope refers to cordage delivered to a ship from a ropeworks.

loblolly boy: an enlisted man assisting the ship's surgeon.

log: a journal accounting for the ship's movements and actions; a section of wood trailed overboard to measure the ship's speed (see knot).

luff: to sail a ship close to the wind (i.e., as nearly into the wind as possible), sometimes as a battle tactic to reduce her speed (luff up).

main deck: the uppermost deck of a ship continuous from stem to stern.

mainmast: the second mast aft on a ship with two or more masts; generally the largest mast.

meridian: noon.

mess: a group of men that eats together, eight in the case of midshipmen.

midshipman: an apprentice naval officer, so named from being quartered amidships.

mizzenmast: the third mast aft on a ship with three or more masts.

nautical mile: a distance equivalent to one minute of the earth's 360-degree circumference at the equator, approximately 6,080 feet or 1.15 statute miles of 5,280 feet.

ordinary seaman: the lowest ranking member of a ship's company (excluding landsmen and boys); with sufficient sea time and knowledge he becomes an able seaman.

orlop: the lowest deck in a warship.

outboard: away from the center of a ship.

purser: the officer in charge of acquiring stores and managing a ship's accounts.

pinnace: a large boat used to convey supplies to the ship or to land shore parties.

punt: a small, square boat used to maintain the exterior of the ship.

quarter: the starboard and larboard quarters are the areas to 45 degrees of either side of dead astern.

Quarter Bill: see Watch, Quarter, and Station Bill.

rake: to cannonade a ship with broadsides at the stern or bow, so that the shot scours the whole length of the decks. The most advantageous battle maneuver at sea, raking delivers maximal fire into an enemy ship at its weakest points while exposing the strongest portion of one's own ship to minimal fire in return.

reef: to reduce the area and thus the drawing power of a sail by tying it off at reef points.

rigging: the lines and associated equipment used to support topside structures and perform their work; rigging generally subdivides into *standing*, which supports and stiffens masts, and *running*, which the crew adjusts to reposition yards, trim sails, etc.

scuttlebutt: the cask from which the daily fresh water ration is issued to the crew.

seine: a fishing net that hangs vertically from floats, with the ends being drawn together to surround the fish.

ship of the line (line-of-battle ship): the most formidable of warships, the smallest of two decks and 74 guns; designed for fleet service in line of battle.

shiver: the fluttering of a sail as it begins to lose drawing power when sailing too close to the direction of the wind.

slops: articles of clothing or bedding sold or issued to sailors.

spar: a stout, usually tapered, pole used in rigging, such as a yard or gaff.

spring: to crack a mast, yard, etc., due to carrying excessive sail in blowing weather, rendering it unsafe for further use without either replacement or temporary repair (called fishing) using wooden splints and iron hoops.

square rigged: A sail arrangement in which most of the sails are suspended from yards carried roughly perpendicular, or "square," to the fore-and-aft center line of the ship; this arrangement maximizes the ship's power at the expense of the maneuverability offered by a fore-and-aft sail plan.

starboard: the right side of a ship when facing forward; to the right.

stays: paired lines of standing rigging leading forward (forestays) or aft (backstays) that support and stiffen the masts; forestays are also used to carry fore-and-aft staysails.

steerage: the area on the berth deck aft of the mainmast where midshipmen eat and sleep.

stem: the solitary vertical framing member attached to the keel at the very bow of the ship and to which the outer and inner planks are affixed.

stern: the after end of a ship.

tack: to work a ship to windward sailing on alternate starboard and larboard tacks (the tack being named for the direction from which the wind strikes the sails); to shift a ship from one tack to the other by putting her bow through the wind.

tierce: a measure of wine usually equivalent to 35 gallons (about 156 litres).

top: a platform built onto a mast, generally at the junction of the lower and upper masts about a third of the way aloft and used to anchor the topmast shrouds. During battle, marksmen are assigned to enlarged tops, called fighting tops, to attack those on the deck of the enemy ship, while others are stationed in the tops to repair damage to rigging.

taffrail: An ornamental railing along the upper edge of the stern.

trim:

ship: to establish a correct balance, especially fore and aft, through proper distribution of ballast and cargo, a responsibility of the sailing master;

sails: to arrange the sails most efficiently for the course being steered.

truck: the highest part of the fore or main mast; the highest part of the mizzen mast is the cap.

'tween decks: the interior of the ship; between the decks.

veer (or wear): to shift a ship from one tack to the other by putting her stern through the wind.

waist: the part of a ship between the forecastle and quarterdeck; roughly amidships.

waister: a sailor or marine assigned at quarters or stations to a position in the waist of the ship, usually to heave on lines.

wardroom: the officers' common room at the rear of the ship used for eating and off-duty activities.

Watch, Quarter, and Station Bill: the master list indicating the individual assignments for every crew member for regular duty (watches), battle (quarters), or special assignments (stations) such as anchoring or working boats.

wear: see veer.

weather: toward the wind; synonymous with windward.

weather gauge: to hold a position to windward of another ship; this gives an advantage of maneuver often critical in battle.

weather side: the side of the ship upon which the wind blows.

whaleboat: a double-ended ship's boat, the most maneuverable and safest in a sea-way.

windward: in the direction from which the wind comes (opposite: lee).

wings: open spaces maintained on the orlop deck along the ship's sides to give carpenters access to plug shot holes during battle.

worm: an iron double claw on a stave used to extract debris from a cannon after firing; the act of doing so.

yard: a long timber centered on a mast and used to carry square sails.

yaw: the starboard or larboard deviation of a ship from her course as she works in a seaway due to the effects of wind or current.

yeoman: a petty officer who serves as a clerk or secretary.

ADDENDA TO THE 1816 EDITION

CONSTITUTION'S EXPLOITS OF 1815

Constitution *remained for most of 1814 blockaded in Boston by the British fleet. On December 17, with the blockading force driven off by winter weather,* Constitution *put to sea. On her last war cruise she would patrol the eastern Atlantic to intercept British ships sailing between Europe and the Caribbean. Here, in February and March 1815* Constitution *would have her final confrontations with the Royal Navy, a fight and a flight.*

Defeat of HMS **Cyane** *and HMS* **Levant**

United States Frigate *Constitution*
At Sea, 23d February 1815

Sir:—On the twentieth of February last, the Island of Madeira bearing W. S. W., distant about sixty leagues [180 nautical miles], we fell in with his Britannic Majesty's *two* ships of war, the *Cyane* and *Levant*, and brought them to action about 6 o'clock in the *evening*, both of which, after a spirited engagement of forty minutes, surrendered to the ship under my command.

Considering the advantages derived by the enemy from having a divided and more active force, as also the *superiority* in the weight and number of their guns, I deem the speedy and decisive result of this action, the strongest assurance which can be given to the Government, that all under my command did their duty, and gallantly supported the reputation of American seamen.

Enclosed you will receive the minutes of the action, and a list of the killed and wounded on board this ship: also, enclosed you will receive for your information, a statement of the actual force of the enemy, and the number killed and wounded on board their ships, as near as could be ascertained.

I have the honor to be, sir, very respectfully,

Your obedient servant,
Charles Stewart

The following is a list of the killed and wounded on board the Constitution in her engagement with the two British sloops of war the *Cyane* and *Levant*; likewise a list of the killed and wounded on board the two latter ships.

Killed on board the *Constitution*—Jno. Fullington OS; Anthony Farrow, marine; William Harral, do.

Wounded—David Quill, quarter master, slightly; James Jackson, S severely; Tobias Fernald, S since dead; B. Thomas, S severely; B Benderford, S slightly; Vincent Marks, S severely; J Lancy, O S. since dead; Thomas Fessenden, O. S since dead; Benjamin Norekroff, Sgt M severely; William Holmes, P. M severely; Patrick Cain, P. M. severely; Andrew Chambers, P M. slightly.

Force of the *Constitution*, thirty-two 24-pounders, 20 32 do./52 guns.

Officers, men and boys 466.

Killed on board the *Cyane* twelve—wounded 26.

Force of the *Cyane* twenty-two 32-pounders, ten eighteen do., 2 12 do./34 guns & 2 brass swivels.

Officers, men and boys 180.

Killed on board the *Levant* 23—wounded 16.

Force of the Levant 18 thirty-two pounders, 2 nine do., 1 twelve do./21 guns.

Officers, men and boys 156.

N.B. that in his action against these two smaller but more maneuverable ships, Captain Stewart used Constitution's *strengths to dictate the terms of the engagement. Both Captain George Douglas of* Levant *and Captain Gordon Falcon of* Cyane *mentioned this in their reports concerning the defeat and capture of their ships.*

Captain Douglas wrote of a point in the encounter found Levant *"leaking considerably" and* Cyane *"apparently unmanageable,"*

the enemy at this time appeared to have suffered but little, in consequence of his being to windward during the action, and keeping at too great a distance to allow our Carronades to do full execution.

Similarly, Captain Falcon recorded:

it was my constant endeavour to close with the Enemy, finding we were too far distant for Carronades, at the same time exposed to the full effect of his long guns, and obtain a position on his quarter, in this however I was only partially successful, as the situation and superior sailing of the enemy's ship enabled him to keep the *Cyane* generally on his broadside, consequently exposed to a heavy fire, from which in the early part of the action the Ship suffered very much in the rigging and latterly in the hull.

At the time of his surrender, Falcon reported of Cyane's *condition*:

the ship, having nearly the whole of the standing and all the running rigging cut, the sails very much shot and torn,—all the lower masts severely wounded, particularly the main and mizzen masts, both of which were tottering, fore yard, fore and mizzen topmasts, gaff and driver boom, main topgallant yard and fore topgallant mast, shot away or severely wounded,—a number of shot in the hull eight or nine between wind and water,—six guns dismounted or otherwise disabled by shot, drawing off bolts, &c.—with a considerable reduction from our strength in killed and wounded.

Mr. Midshipman Pardon Whipple of USS Constitution *gives a similar account of the damages sustained by HMS* Cyane *and HMS* Levant *in their encounter with his ship*:

Our boats were employed during the principal part of the night removing prisoners and repairing damages, so that before morning we were enabled to make some sail on the prizes. The *Constitution* was so little injured and the damages repaired with such expedition that in one hour after the close of this action she was ready for another.

Being in one of the boats that night I had a good opportunity of estimating the injury we had done them, which was very considerable during so short a time that we were engaged.

I first had charge of a boat removing prisoners from the *Cyane* and afterwards from the *Levant*. I remained on board of that vessel three days. The decks of both vessels were literally covered with dead and wounded. . . . Their spars and riggings were much cut to pieces, particularly the *Levant*'s whose mizzen mast and all the appendant spars were wounded or carried away. . . .

This being the first action I was ever in, you can imagine to yourself what my feelings [were] to hear the horrid groans of the wounded and dying, and the scene that presented itself the next morning at daylight on board of the *Levant*. The quarter deck seemed to have the appearance of a slaughter house, the wheel having been carried away by a shot, killed and wounded all round it.

The mizzen mast for several feet was covered with brains and blood; teeth, pieces of bones, fingers and large pieces of flesh were picked up from off the deck. It was a long time before I could familiarize myself to these and [if] possible more horrid scenes that I had witnessed.

Having demonstrated her superiority in arms, Constitution *escorted her prizes to the neutral Portuguese Porto Praya in the Cape Verde Islands to arrange for parole of her prisoners. There, for the second time during the war,* Constitution *would be required to demonstrate her superiority in sailing to avoid being captured herself.* Constitution's *log from 11 March 1815 reads:*

Remarks on board U.S. frigate *Constitution*,
Charles Stewart Esqr. Commander.

At 8 A.M. . . . . Sent Captains Douglass & Falcon on shore; at meridian they returned and said they had arranged for the cartel. Sent a boat and an officer to bring the English brig under our stern to have her convenient to provision &c. Made preparations for supplying the cartels from the prizes. [The afternoon] commences with fresh breezes and thick foggy weather. At 0 h. 5 m. P.M. discovered a large ship through the fog standing in for Port Praya. At 0 h. 8 m. discovered two other large ships astern of her also standing in for the Port. From their general appearance supposed them to be one of the enemy's squadrons, and from the little respect hitherto paid by them to neutral waters I deemed it most prudent to put to sea. The signal was immediately made to the *Cyane* and *Levant* to get under weigh. At 0 h. 12 m. with our topsails set we cut our cable and got under weigh, when the Portuguese opened a fire upon us from several of their batteries on shore. The prize ships followed our motions and stood out of the harbour of Port Praya close under East point passing the Enemy's squadron about gun shot to windward of them; crossed our topgallant yards and set foresail, mainsail, spanker, flying-jib and topgallant sails. The enemy seeing us under way tacked ship and made all sail in chase of us. As far as we could judge of their rate from the thickness of the weather, supposed them to be two ships of the line and one frigate. At 0 h. 30 m. cut away the boats towing astern, first cutter and gig. At 1 P.M. we found our sailing about equal with the ship on our lee quarter, but the frigate luffing up and gaining our wake and rather dropping astern of us. The *Cyane* dropping fast astern and to leeward and the frigate gaining on her fast I found it would be impossible to save her if she continued on the same course without having the *Constitution* brought to action by their whole force; I made the signal at 1 h. 10 m. to her to tack which was complied with. This maneuver I conceived would detach one of the enemy's ships in pursuit of her, while at the same time from her position she would be enabled to reach the anchorage at Port Praya before the detached ship would come up with her; but if they did not tack after her it would afford her an opportunity to double their rear and make her escape before the wind. They all continued in full chase of the *Levant* and this ship, the ship on our lee quarter firing her broadside by divisions the shot falling short of us. At 3, having dropped the *Levant* considerably, her situation became from the position of the Enemy's frigate similar to the *Cyane*'s. It now became necessary to separate also from the *Levant* or risk this ship being brought to action to cover her; the signal was accordingly made at 3 h. 5 m. P.M for her to tack which was complied with. At 3 h. 12 m. the whole of the enemy's squadron tacked in pursuit of the *Levant* and gave over the pursuit of this ship. This sacrifice of the *Levant* became necessary for the preservation of the *Constitution*. Set the royals and kept large from the wind. Sailing Master Hixon, Midshipman Varnum, one boatswain's mate and twelve men, who were absent on duty in the 5th cutter to bring the cartel brig under our stern were left on board the *Levant,* which ship they reached before she cut. Surgeon's Mate Johnson, with the sailmaker and his mate, were likewise on board the *Levant.*

A CELEBRATED SHIP

Old Ironsides won her latest fight and made her latest escape three days and three weeks, respectively, after the U.S. Senate ratified the Treaty of Ghent ending the recent war against the United Kingdom. Since her victories over HMS Cyane *and HMS* Levant *came within the grace period stipulated in the treaty, they were legitimate prizes of war. USS* Constitution *returned home, three months after peace was declared, a celebrated ship.*

The *Constitution* Arrived

We announce with pleasure the safe arrival at [New York] of the United States frigate CONSTITUTION of 44 guns, Charles Stewart, Esq. commander, from a cruise of about five months. The career of "Old Ironsides" has been brilliant and fortunate almost beyond example. Twice . . . has she been encompassed by British squadrons, but by judicious and skillful management gave them both times the slip. At the commencement of the war she struck the first blow—followed it up in the middle by another equally severe—and at its conclusion by a third, capturing two sloops of war.—*Columbian Centinel* (Boston), 20 May 1815.

A poetic tribute to Old Ironsides sung to "Yankee Doodle," 1815:

> Then raise amain, the joyful strain,
> For well she has deserv'd it,
> Who brought the foe so often low,
> Cheer'd freedom's heart and nerv'd it;
> Long may she ride, our navy's pride,
> And spur to resolution;
> And seamen boast, and landsmen toast,
> The Frigate *Constitution*.

RESOLUTION,
Requesting the President of the United States,
to present medals to Captain Charles Stewart
and the officers of the frigate *Constitution*.

January 10, 1816.

Resolved, by the Senate and House of Representatives of the United States of America, in Congress assembled, That the President of the United States be, and he is hereby requested to present to captain Charles Stewart, of the frigate *Constitution*, a gold medal with suitable emblems and devices, and a silver medal with suitable emblems and devices, to each commissioned officer of the said frigate, in testimony of the high sense entertained by Congress, of the gallantry, good conduct, and services of captain Stewart, his officers and crew, in the capture of the British vessels of war, the *Cyane* and the *Levant*, after a brave and skillful combat.

USS Constitution's *victories and the disposition of said vessels during the late war recently concluded*:

Date	Vessel Name	Vessel Type	Disposition
10 August 1812	Lady Warren	Brig	Destroyed
11 August 1812	Adeona	Brig	Destroyed
15 August 1812	Adelaide	Brig	Recaptured
19 August 1812	HMS Guerriere	Frigate	Destroyed
8 November 1812	South Carolina	American Brig	Restored to Owner
29 December 1812	HMS Java	Frigate	Destroyed
14 February 1814	Lovely Ann	Ship	Cartel (Truce Ship)
14 February 1814	HMS Pictou	Schooner	Destroyed
17 February 1814	Phoenix	Schooner	Destroyed
19 February 1814	Catherine	Brig	Destroyed
24 December 1814	Lord Nelson	Brig	Destroyed
16 February 1815	Susannah	Ship	Prize
20 February 1815	HMS Cyane	Light Frigate	Prize
20 February 1815	HMS Levant	Corvette	Recaptured

Our National Ship, the *Constitution*, is once more arrived.

Let us keep "*Old Iron Sides*" at home. She has, literally, become a *Nation's* Ship, and should be preserved. Not as a "sheer hulk, in ordinary" (for she is no *ordinary* vessel); but, in honorable pomp, as a glorious Monument of her own, and our other Naval Victories.

She has "*done her duty*"; and we can therefore *afford* to preserve her from future dangers.

Let a dry dock, such as are used in Holland, and other parts of Europe, be contracted for her reception, at the Metropolis of the United States. Let a suitable and appropriate building be erected over her, to secure her from the weather; and other measures used to preserve her from decay: that our children, and children's children, may view this stately monument of our National Triumphs.

The decks of this noble Ship have witnessed peculiarly striking instances of superiority and success over her enemies.—When in battle, the skill and courage of her officers and crew, have invariably brought her victory; and when pursued by a superior force (frequently happening) the superior seamanship of her different commanders has completely baffled the efforts of her foes, and preserved her for new and splendid triumphs!

"She has done her duty"; she has done ENOUGH!

Let us preserve her as a precious model, and example for future imitations of her illustrious performances!

—*The National Intelligencer* (Washington, D.C.), 23 May 1815

REFERENCES

INTRODUCTION

Navy in the Federal Era:

Charles E. Brodine, Jr., Michael J. Crawford, and Christine F. Hughes, *Interpreting Old Ironsides: An Illustrated Guide to USS Constitution* (Washington, D.C.: Naval Historical Center, 2007), 36.

Barbary Corsairs:

Brodine, *Interpreting Old Ironsides*, 36-7.

Quasi-War:

Brodine, *Interpreting Old Ironsides*, 34-6;

Arthur M. Schlesinger, Jr., *Almanac of American History* (NY: Perigee Books, 1983), 169-75.

Early Service *of Constitution*:

Tyrone G. Martin, *A Most Fortunate Ship: A Narrative History of Old Ironsides* (Annapolis: USNI Press, 1997), 23-66, 82-126;

Schlesinger, *Almanac*, 179-82;

Register of Officer Personnel United States Navy and Marine Corps and Ships' Data 1801-1807 (Washington, D.C.: USGPO, 1945), 53, 70-1.

Napoleonic Wars:

Schlesinger, *Almanac*, 182-6;

Brodine, *Interpreting Old Ironsides*, 67-70.

"Free Trade and Sailors' Rights":

Schlesinger, *Almanac*, 176-93;

Brodine, *Interpreting Old Ironsides*, 67-70. The U.S. declaration of war came two days after the British government had rescinded its Orders in Council, less to satisfy the Americans than to mollify British merchants adversely affected by the U.S. embargoes.

Mr. Madison's War:

A New England Farmer, *Mr. Madison's War* (Boston: Russell & Cutler, 1812);

Schlesinger, *Almanac*, 192-8.

Peace Feelers:

Schlesinger, *Almanac*, 194-202.

U.S. Navy in War of 1812:

"Exhibit shewing the number of Vessels of War of the United States, now in active ser-
    vice; their names, rates, and stations for the winter" from: *Report of the Committee to
    Whom Was Referred so much of the President's message as Relates to the Naval Establish-
    ment. December 17, 1811* (Washington City: A & G Way, Printers, 1811);

Brodine, *Interpreting Old Ironsides*, 41-2, 48-9;

Schlesinger, *Almanac*, 193-8.

The Year 1814:

"List of the Naval Force of the United States [on the Atlantic, 4 March 1814]" in: *Docu-
    ments from the Secretary of the Navy Related to the Navy of the United States, March
    18, 1814* (Washington City: Roger C. Weightman, 1814);

Brodine, *Interpreting Old Ironsides*, 42, 45-8, 67-70, 72;

Schlesinger, *Almanac*, 198-201.

USS *Constitution* in 1814:

Brodine, *Interpreting Old Ironsides*, 28-9;

Schlesinger, *Almanac*, 138-9.

Midshipmen:

Brodine, *Interpreting Old Ironsides*, 11;

Literacy:

Matthew Brenckle, Lauren McCormack, and Sarah Watkins, *Men of Iron: USS* Constitu-
    tion*'s War of 1812 Crew* ([Boston]: USS *Constitution* Museum, 2012), 31;

Profiled:

Christopher McKee, A Gentlemanly and Honorable Profession: The Creation of the U.S.
    Naval Officer Corps, 1794-1815 (Annapolis, USNI Press, 1991), 40, 64, 66, 68-71,
    89-94.

Constitution's Last War Patrol:

Brodine, *Interpreting Old Ironsides*, 139-40.

Constitution's Subsequent Service:

Pope's Visit:

Microfilm M1030, Roll 1, Record Group 24, Logbooks and Journals of the USS *Constitu-
    tion*, 1798-1934, U.S. National Archives and Records Administration, Washington,
    D.C. (hereafter: NARA);

Circumnavigation:
Microfilm M1030, Roll 1; Martin, *Most Fortunate Ship*, 266-89;

1920s Survey and Reconstruction:
Commandant, U.S. Navy Yard, Boston, "U.S. Frigate *Constitution* (IX21) Research Memorandum, Nov 27 1931," A/Reports, Cat. #1302.39, USS *Constitution* Museum Library and Archives, Charlestown, MA (hereafter: *Constitution* Archives);

National Cruise:
Program for "General Public Reception on Sunday Afternoon, May 13, 1934 to the USS *Constitution* celebrating her return home to Boston from the National Cruise," Cat. #1925.1, *Constitution* Archives;

1954 Act:
Booklet H.R. 8247, An Act to Provide for the Restoration and Maintenance of the United States Ship *Constitution*," 4 May 1954, Cat. #1493.3, *Constitution* Archives;

Bicentennial Events:
Martin, *Most Fortunate Ship*, 362-8.

Events post-1995:
USS *Constitution* website: www.navy.mil/local/constitution/visitors.asp.

And a Very Happy Ending:
Rush-Bagot Agreement:
Exchange of Notes Relative to Naval Forces on the American Lakes, signed at Washington April 28 and 29, 1817;
Jeffrey A. Larsen and James M. Smith, *Historical Dictionary of Arms Control and Disarmament* (Lanham, MD: Scarecrow Press, 2005), 187;
Michael A. Hennessy and B. J. C. McKercher (eds.), *War in the Twentieth Century: Reflections at Century's End* (Westport, CT: Praeger, 2003), 60.

MISTER MIDSHIPMAN

War Message:
Secretary Hamilton, 19 June 1812, Microfilm M977, General Orders and Circulars, 1798-1862, NARA.

Oath:
Oath of Allegiance of Alexander Eskridge, 22 Feb. 1812, from Brodine, *Interpreting Old Ironsides*, 13.

Duties of:

"Of the Duties of Midshipmen," Naval Regulations Issued by Command of the President of the
United States of America, [1814], reprinted in Brodine, *Interpreting Old Ironsides*, 120-9.

Keeping of Journals:

Navy Regulations of 1814.

Dress:

U.S. Navy Uniform Regulations, 1814, reprinted in Brodine, *Interpreting Old Ironsides*, 131-2.

Quarters and Mess:

Naval History and Heritage Command Detachment Boston, "USS *Constitution* Plans:
For Model Ship Builders and General Researchers" (CD. Washington, D.C.: Naval
Heritage Foundation, 2012);

"Plans of the Frigate United States, Decks, &c." (illustration), NARA;

Tyrone G. Martin, *A Signal Honor: The Men of* Constitution (Chapel Hill, NC: Tryon
Publishing Co., 2003), 46.

Deportment:

Regulations from *An Act for the Better Government of the Navy of the United States* (23
Apr. 1800), (Sixth Congress, Sess. 1., Ch. 33. 1800);

Captain Edward Preble, "Internal Rules and Regulations for U.S. Frigate Constitution, 1803-
1804," in: *U.S. Navy, Naval Documents Related to the United States Wars with the Barbary
Powers* (6 vols., Washington, D.C.: USGPO, 1939), v. III, 32-41 (hereafter: Preble's Orders).
Although Preble issued his orders eleven years before the date chosen for this volume, they
are mostly repeated, often word for word, in Captain John Rogers orders for *Constitution*
of 1809 (reprinted in Martin, *Singular Honor*, 69-93). It seems reasonable to conclude that
something very like them would have been in force aboard *Constitution* in 1814.

Admonishment:

Edmund March Blunt, *Seamanship Both in Theory and Practice* (New York: Edmund M.
Blunt, 1813), 207.

THE FRIGATE *CONSTITUTION*

Formation of the Navy:

U.S. Constitution, Art. 1., Sec. 8; "An Act to Provide a Naval Armament," Third Congress.
Sess. I. Ch. 12. 1794.

Design of Class:

Brodine, *Interpreting Old Ironsides*, 33-4.

Construction:

Brodine, *Interpreting Old Ironsides*, 6-7;

"Detailed estimate of the expense of building and equipping a 44-gun frigate in 1798,"
in: Commodore George Henry Preble, U.S.N., "History of the Boston Navy Yard in
Charlestown, Mass. from 1797 to 1875," U.S. Navy Department, 1875, 28-9;

Letter from the Secretary of War Regarding Military Forces and Progress on the Frigates,
14 Dec. 1795, Cat. #2183.2, *Constitution* Archives;

Letter from the Secretary of War, Transmitting Sundry Statements Relative to the
Frigates *United States*, *Constitution* and *Constellation*, June 17th, 1797, Cat. #2073.1,
Constitution Archives;

Claghorne to Sec. of War in U.S. Navy Bureau of Construction and Repair, letters pertaining to George Claghorne, 15 Mar. 1926, Cat. #1901.1, *Constitution* Archives.

Entered Service:

Brodine, *Interpreting Old Ironsides*, 143;

Total Cost:

Register of Officer Personnel United States Navy and Marine Corps and Ships' Data 1801-1807
(Wash: USGPO, 1945), 70.

Dimensions, Characteristics, and Features:

Pamphlet, Helen A. Faber, "The United States Frigate *Constitution* "Old Ironsides," 1933,
Cat. #1928.1, *Constitution* Archives;

Typed Sheet of Ship's Characteristics, A/Reports, United States Frigate *Constitution*
Historical, Cat. #1917.4, *Constitution* Archives;

Hand-written Note Stating Ship's Characteristics, 12 July. 1945, *Constitution* Restoration,
1927-31 (Correspondence re, 1927-48), Cat. #740.6, *Constitution* Archives;

Register of Officer Personnel, 70-1;

Brodine, *Interpreting Old Ironsides*, 5, 143;

Navy Historical Center, Curator's Office, Washington, D.C., (www.navy.mil/navydata/
fact_display.asp?cid=4200&tid=100&ct=4).

Slight differences occurred with some statistics; I used either the consensus figure or
what I took to be the best source.

MANNING *CONSTITUTION*

Ship's Company:

Martin, *Signal Honor*, 95-6.

Lieutenants:
Navy Regulations of 1814, Brodine, *Interpreting Old Ironsides*, 123-4.

The Ship's Divisions:
Blunt, *Seamanship*, 242;
Preble's orders re divisions, 2 Sep. 1803, *Wars with the Barbary Powers*, 7.

The Ship's Marines:
Preble's Orders, *Wars with the Barbary Powers*, 39, 41.

Pay and Provisions:
Tables were calculated using:
*Letter from the Secretary of War Transmitting Sundry Statements Relative to the Frigates
    United States,* Constitution *and* Constellation, *June 17th, 1797* (Philadelphia: Zacha-
    riah Poulson, Junior, 1797);

Allotments:
An Act for the better government of the Navy of the United States. April 23, 1800 (Sixth
    Congress, Sess. 1., Ch. 33. 1800.);
Allotment Form: A/Certificates, Cat. #2087.1, *Constitution Archives*;

Rations:
"An Act to Provide a Naval Armament," 1794.

Ship's Discipline:
Brodine, *Interpreting Old Ironsides*, 19-21;
Blunt, *Seamanship*, 207.

MATTERS MEDICAL

Injury and Disease:
Brodine, *Interpreting Old Ironsides*, 59-61.

Medical Text:
Plain Remarks on the Accidents and Diseases which Occur Most Frequently At Sea (Boston:
    Howe's Sheet Anchor Press, 1846), 21-2, 28-31.

Cleanliness at Sea:
Blunt, *Seamanship*, 208-9;
Surgeon's Remarks:
Journal Kept on board the Frigate Constitution *1812 by Amos A. Evans, Surgeon U.S.N.*
    (reprint, Concord, MA: Bankers Lithograph Co., 1967), 470.

Incidents:

Boatswain Shot:

Constitution's log, 2 Mar. 1799;

Man Overboard:

Evan, *Journal*, 166.

Even in war, disease and accident were dangers to be feared as much as combat. Of
   Constitution's fifty-two recorded wartime deaths, twenty-seven were due to combat,
   five to accident, and twenty to disease (Brodine, *Interpreting Old Ironsides*, 61).

PROVISIONING *CONSTITUTION*

Regulations Respecting:

Blunt, *Seamanship*, 179-80;

U.S. Naval Regulations, 1814, Brodine, *Interpreting Old Ironsides*, 126-7.

Inventory of:

"Inventory of an Unidentified Frigate, 1799," A/Reports, Cat. #1989.1, *Constitution* Archives.

Resupply and Grog Ration:

Brodine, *Interpreting Old Ironsides*, 17-8;

Eating fish caught, acquiring fruits at island, attempting to catch rainwater at sea:

Evans, Journal, 469-74.

Regarding Slops:

U.S. Navy Regulations, 1814, Brodine, *Interpreting Old Ironsides*, 127;

Preble's Orders, 7, 32, 33, 36.

Ship's Stores:

Cooper's stores:

"Inventory of an Unidentified Frigate, 1799," A/Reports, Cat. #1989.1, *Constitution*
   Archives.

Letter concerning stationary: A/Ephemera, Cat. #2002.1, *Constitution* Archive.

SAILING *CONSTITUTION*

Shipboard Routine:

Blunt, *Seamanship*, 219.

Maneuvering the Ship:

Blunt, *Seamanship*, 105-6, 110, 226;

Brodine, *Interpreting Old Ironsides*, 12; Preble's Orders, 6-7.

Captain's Orders:
Preble's Orders, 32-8.

Signaling:
Blunt, *Seamanship*, 148, 152, 158, 165.

Fire at Sea:
Evans, *Journal*, 166;
Blunt, *Seamanship*, 232-4.

Boats and Boatmanship:
"U.S. Frigate *Constitution* (IX21) Research Memorandum, Nov 27 1931" *Constitution*
    Archives;
Rowing Commands:
Derived from San Francisco National Maritime Museum Park Association, "Lesson
    Plans, Dory Procedures," www.maritime.org/edu/crew-packets.htm; "Pulling Boat
    Oar Commands," HMS *Richmond* . . . Living History, 1775-1783, www.hmsrich-
    mond.org/oarcmd.htm.

TO FIGHT THE SHIP

Quarter Bill:
Blunt, *Seamanship*, 225, 245-6.

***Constitution*'s Armament:**
Brodine, *Interpreting Old Ironsides*, 7-10, 143.

Clearing for Action:
Brodine, *Interpreting Old Ironsides*, 51;
Blunt, *Seamanship*, 244-5, 247;
"A Guide to and History of U.S. Frigate *Constitution*," (Worcester, MA: R. E. McTyre,
    1931), 29-30.

Engaging the Enemy:
Blunt, *Seamanship*, 195, 252;
Preble's Orders, 38;
Brodine, *Interpreting Old Ironsides*, 56, 62-5.

Boarding:
Brodine, *Interpreting Old Ironsides*, 15, 122;
U.S. Naval Regulations of 1814, Brodine, *Interpreting Old Ironsides*, 122.
Blunt, *Seamanship*, 128, 251-2, 255-6.

OF PRIZES AND PENSIONS

Rules Governing Prizes:

Norma Adams Price (ed.), *Letters from Old Ironsides, 1813-1815: Written by Pardon Mawney Whipple, USN,* Tempe, AZ: Beverly-Merriam Press, 2006, 17;

An Act for the Better Government of the Navy of the United States, April 23, 1800 (Sixth Congress, Sess. 1., Ch. 33. 1800);

Secretary of the Navy to the Chairman of the House Naval Committee, in William S. Dudley (ed.), *The Naval War of 1812: A Documentary History, Volume 1 1812* (Washington, D.C.: USGPO, 1985), 578-9;

"Message from the President of the United States, Transmitting a Letter from Cap. Bainbridge . . . February 22d, 1813," (Washington City: Roger C. Weightman, 1813);

Laws of the United States in Relation to Navy and Marine Corps; To the Close of the Second Session of the Twenty-Sixth Congress (Washington: J. and G. S. Gideon, 1843), 175-6.

Prize Estimate Table:

Calculated using the rules stipulated in *An Act For the Better Government of the Navy* (1800), and the highest paid person in the category from "Estimate of the pay and rations of the officers and crew of a ship of war of 74 guns for twelve months—650 men," 1811. The 1814 crew numbers used appear in Martin, *Signal Honor,* 95-6. This estimate of men aboard might be slightly low; in his action report, Stewart listed the crew as numbering 466. This would reduce the size of the individual prizes slightly, mostly among the able and ordinary seamen.

Pensions Awarded for Naval Service:

System:

An Act for the Better Government of the Navy of the United States (1800);

Killed:

Blunt, *Seamanship,* 278;

Disabled:

From *An Act Providing a Naval Armament,* 1 July 1797.

"OLD IRONSIDES": HER EXPLOITS OF 1812

The Declaration:

"Report of the Committee of the Senate of Massachusetts . . . The Act Declaring War. . . . June 26, 1812" (Boston: Adams, Rhoades and Co., Printers, 1812), A/Pamphlets, Catalog #1370.12, *Constitution* Archives (corrected in minor details with "Acts of the Twelfth Congress of the United States," Session 1, Ch. 102, p. 755).

The Chase:

Weekly Messenger (Boston), 31 July 1812;

In his report to Navy Secretary Hamilton, Hull recorded the dates of the engagement
mentioned in the paper as 17 and 19 July (Microfilm M125, Roll 24, Letters Received by
the Secretary of the Navy from Captains, 1805-61 (hereafter: "Captains' Letters") NARA.

Defeat of HMS *Guerriere*:

Salem Gazette, 1 Sep. 1812; Hull to Navy Secretary Hamilton, 28 Aug. 12, Microfilm
M125, Roll 24, "Captains' Letters," NARA; Dacres to Sawyer, Dudley, *Naval War of
1812*, 243-5.

Defeat of HMS *Java*:

Constitution's log, 29 Dec. 1812, M1030, Roll 1, NARA; Chads to Croker, 31 Dec. 1812,
Dudley, *Naval War of 1812*, 646-8.

Heroes of USS *Constitution*:

Niles Weekly Register (Washington, D.C.), 12 Sep. 1812; Lt. Contee to Lewis Bush, 13 Sep.
1812, Cat. #1649.1, Constitution Archive; Evans, *Journal*, 479.

British Reaction to the American Victories:

The Times (London), 20 Mar. 1813.

A NAVAL AND NAUTICAL GLOSSARY

Sources Included:

Martin, *Most Fortunate Ship*, 381-7;

Blunt, *Seamanship*, 9-26;

John V. Noel, Jr., and Edward L. Beach, *Naval Terms Dictionary*, 4th ed. (Annapolis: USNI
Press, 1978).

"Port" replaced "larboard" officially in a Navy Department General Order of 18 February
1846 (Microfilm M977: Navy Department General Orders and Circulars, 1798-1862,
NARA).

ADDENDA

Defeat of HMS *Cyane* and HMS *Levant*:

Stewart to the secretary of the navy, 23 Feb. 1815, M125, Roll 44, NARA (also in
Biographical Sketch and Services of Commodore Charles Stewart (Philadelphia: J.
Harding, 1838), 23.

Casualties:

Weekly Messenger (Boston), 19 May 1815;

Douglas to Croker, 22 Feb. 1815, and Falcon to Douglas, 22 Feb. 1815, in Brodine, *Interpreting Old Ironsides*, 110-2;

Price, *Letters from Old Ironsides*, 20-1;

"Remarks on board U.S. frigate *Constitution*," 11, Mar. 1815, M1030, Reel 1, NARA.

A Celebrated Ship:

"The *Constitution* Arrived," *Columbian Centinel* (Boston), 20 May 1815.

Yankee Doodle Poem:

"'Old Ironsides' Her History 'She Saved the Nation: Let Us Save Her!' Patriotic Rally, Madison Square Garden, June 5, 1926," A/Pamphlets, Cat. #967.1, *Constitution* Archives;

"Resolution Requesting the President of the United States to Present Medals to Captain Charles Stewart and the Officers of the Frigate Constitution, 1816," A/Pamphlets, Cat. #1433.1, *Constitution* Archives.

Captures, 1812-15:

Condensed from a similar list in Brodine, *Interpreting Old Ironsides*, 141;

"Our National Ship," *National Intelligencer* (Washington, D.C.), 23 May 1815.

GENERAL

John Harland, *Seamanship in the Age of Sail* (1984; reprint: Annapolis: USNI Press, 2000).

Andrew Lambert, *War at Sea in the Age of Sail* (2000; reprint: NY: Smithsonian Books, 2005).

Karl Heinz Marquardt, *Anatomy of the Ship: The 44-Gun Frigate USS* Constitution *"Old Ironsides,"* (Annapolis: USNI Press, 2005).

Tyrone G. Martin, *Creating a Legend* (Chapel Hill, NC: Tryon Publishing Co., 1997).

Tyrone G. Martin (ed.), *The USS* Constitution's *Finest Fight, 1815: The Journal of Acting Chaplain Assheton Humphreys, U.S. Navy* (Mount Pleasant, SC: Nautical and Aviation Publishing Co. of America, 2000).

ACKNOWLEDGMENTS

My thanks to Bloomsbury Publishing and editor Lisa Thomas for permitting me to undertake this interesting project. My thanks also to Margherita Desy, historian, Naval History and Heritage Command, Boston; Kate Monea, archivist, USS *Constitution* Museum; and Matthew Brenckle, historian, USS *Constitution* Museum. Their insights have significantly improved it. And my thanks to Barbara Clements, my wife, for proofreading the text, helping to create the glossary, and running a happy ship. The errors remaining are all mine. My limits notwithstanding, I hope that you will enjoy this Conway Pocket Manual and encourage you to explore others in the series.

INDEX

U. S. FRIGATE